THE STORY OF LOWESTOFT LIFEBOATS

Part 3 1924 – 1968

by Stuart Jones BA

Also still available in the same series:

The Story of Lowestoft Lifeboats Part 1, 1801 – 1876
(First published by the Port of Lowestoft Research Society 1973)
(Second Edition published by Lowestoft Libraries 1975,
ISBN 0 904328 00 7, and reprinted 2001)

The Story of Lowestoft Lifeboats Part 2, 1877 – 1924
(First published by the Port of Lowestoft Research Society 1996,
ISBN 0 9505311 3 8)

© **Stuart Jones BA 2008**
All rights reserved. No part of this publication may be reproduced, stored in a retrieval system or transmitted in any form or by any means, electronic, mechanical, photocopying, recording or otherwise, without prior permission from the Port of Lowestoft Research Society.

PLRS008/700

ISBN 978 0 9505311 4 4

British Library Cataloguing in Publication Data.
A catalogue record for this book is available from the British Library.

Published by: THE PORT OF LOWESTOFT RESEARCH SOCIETY
LOWESTOFT, SUFFOLK

Printed by: Tyndale Press Ltd, Lowestoft, Suffolk

CONTENTS

Introduction

Chapter 1 :   Coxswain Spurgeon 1924 – 47

1.1 His early years	1
1.2 The 1930's	13
1.3 The Second World War and after	30
Services	40

Chapter 2 :   Coxswain Burgess 1947 – 68

2.1 Mid 20th century	48
2.2 The lifeboat in the 1960's	59
Services	73

Appendix 1 : Coxswains	81
Appendix 2 : Lifeboat details	83

Index to Lifeboats	87
Index to Vessels	88
Index to People	97

Cover picture : RNLB *Michael Stephens* (ON 838) at Gorleston-on-Sea in connection with the Review of the Herring Fleet in October 1957. This lifeboat served at Lowestoft from 1939 until 1963.
(PLRS collection)

INTRODUCTION

The Port of Lowestoft Research Society has already published Parts 1 and 2 of The Story of Lowestoft Lifeboats in 1974 and 1996 respectively.

I have been asked to continue the series. I should say at once, that without the meticulous notes left by our late Chairman Jack Mitchley the task would have been immeasurably more difficult.

I have ended this volume in 1968 to coincide with the retirement of Coxswain Harry Burgess.

A number of people have been of great help to me in this endeavour. Notably, Bob Collis the Aviation Historian, Jeff Morris the Archivist of the Lifeboat Enthusiasts Society, Mike Chapman the Operations Manager of the Lowestoft Lifeboat station and several members of the PLRS. During early researches for this book I was very privileged to be able to see Albert Spurgeon's Log Books and to be able to make notes from them. My thanks go to his family for this. The Lowestoft Maritime Museum have allowed reproduction of a portrait of Albert Spurgeon which is in their collection, for which I am grateful.
The majority of images in this volume come from the extensive collection of the PLRS which has been built up over many years.

These last 10 to 15 years have seen an exponential growth in the study of family history. To facilitate people with that interest I have included an index of people mentioned in the text in addition to the usual indices.

Figure 1, the location chart, is based on Admiralty chart 1504 by permission of the Controller of Her Majesty's Stationery Office and the UK Hydrographic Office.

Any errors or omissions that there may be are entirely mine. However, I sincerely hope that this volume will be a fitting continuation of the story of the illustrious lifeboatmen of the Lowestoft Station.

SRJ
2008

Figure 1.

Sandbanks off Lowestoft and Great Yarmouth

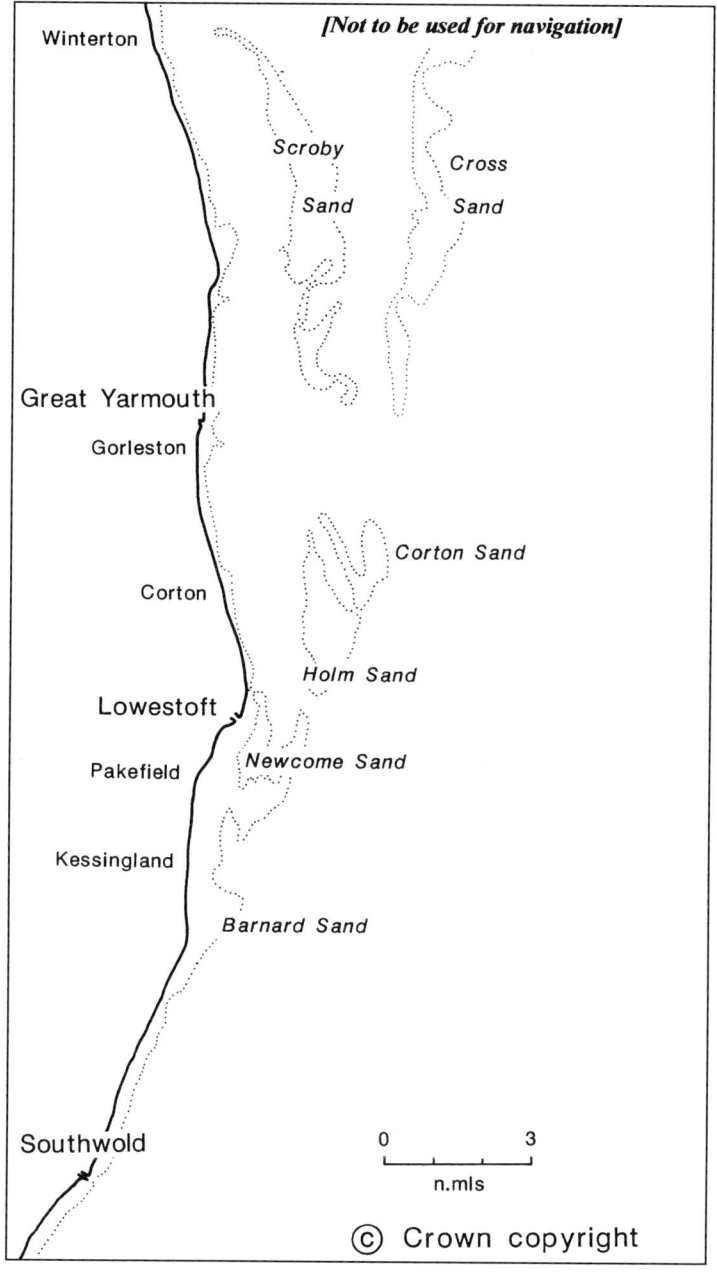

CHAPTER 1

Coxswain Spurgeon 1924 – 1947

1.1 - His early years

Readers of Part 2 of this story will know that Coxswain John Swan retired in June 1924 after serving in the Lowestoft crew for over 50 years. His successor, chosen by a ballot of the members of the Old and New Beach Companies, was Albert Spurgeon who had been a crew member since 1902 and was eventually to serve for a total of 45 years.

His first service came in the following year when on 29th January 1925 the Grimsby steamer **Taunton** ran aground on the Holm Sand, in a moderate southerly gale. Spurgeon took the *Agnes Cross* out at about 9.0am and the Captain of the casualty asked him to stand by. Eventually the wind and sea increased and the lifeboat had to move away and lie to her anchor near by. By four o'clock the vessel was being moved by two tugs, **Lowestoft** from that port and **George Jewson** from Great Yarmouth and so the lifeboat returned to her station. The **Taunton** was eventually refloated at midnight on a higher than usual tide.

A few weeks later the *Lowestoft Journal* of 21st March 1925 reported that Miss Burrows, a sister of Mrs Agnes Cross, had presented the lifeboat with a searchlight, dynamo, riding lights and a klaxon.

During April 1925 a service took place to an airship!! On the 16th the airship **R33** broke from her mooring at Pulham in Norfolk in a severe WSW gale and was blown out over the North Sea. After consultation with the Hon. Secretary and the Coastguard, Coxswain Spurgeon launched the *Agnes Cross* and in company with **HMS Godetia**, the fishery cruiser, followed the drifting airship. After about 14 miles it was decided that the *Agnes Cross* should return to Lowestoft while the warship continued the chase. In winds of 60mph it must have been an uncomfortable passage back for the lifeboat. Letters of appreciation were received by the Institution from the Air Ministry which referred to the flight of **R33** over the North Sea. I suppose it was a flight, if unintended, and the airship did get back to Pulham under the command of Flt.Lt. R S Booth AFC.

Around lunchtime on Saturday 27th June 1925 the smack LT976 **Irene** sailed for the fishing grounds in a strong NE wind with heavy seas. Just north of the harbour mouth she missed stays and was thrown onto the North beach. By 3.0pm she was rolling heavily with seas breaking aboard. The lifeboat crew mustered and by the time *Agnes Cross* got to the smack she had knocked off into shallow water and so the lifeboat lay close by in deeper water. By 4.0pm the **Irene** had filled. The crew shouted for the lifeboat which got alongside and rescued the four men. By early the following week the smack had become a total loss.

Later, *Lloyds List* reported that she had been salvaged and so the Institution sent down an enquiry as to whether she might need the services of the lifeboat again in the future since she and her crew had earlier been rescued by the lifeboat in 1924. Mr GEC Fisher, the acting Station Secretary, replied that he had been to look at the wreck and that there was no possibility of her requiring further services.

By this time Mr R Wollaston-Seago had retired as Station Secretary and Mr Fisher acted as Secretary until towards the end of the year when Mr Sydney Taylor, a local Solicitor, was appointed.

During the autumn herring fishery the Fraserburgh motor drifter FR917 **Diligence** went ashore near Hamilton Road on Tuesday 20th October 1925. Coxswain Spurgeon could not get the *Agnes Cross* near enough in the shallow water and a SE gale and so the line-throwing gun was used for the first time. The lines however, washed off before the crew of the drifter could get to them. The Coastguard life-saving apparatus team fired lines from the shore and successfully took off the crew of eight. Following the Coxswain's report of this service the Chief Inspector of Lifeboats wrote saying he was well pleased with the performance of the gun even if the LSA team had rescued the men in the end.

During the afternoon of Wednesday 25th November 1925 a severe storm began to cause flooding in the beach village area of Lowestoft. In particular Albert Spurgeon could see that the passage to the lifeboat via Hamilton Road would soon become impassable. In company with the Motor Mechanic he made his way round Hamilton Dock dodging heavy seas breaking over the wall. Later they were joined at the lifeboat by Bowman Jack Rose and his brother George. Their fears proved correct, for by 6.0pm it was impossible to get to the lifeboat by land and the dock wall was breached in two places. A while later a motor boat came across the dock and took them off for a meal and a change of clothing. They then returned to the lifeboat, where they stayed until 11.0pm. In his report, Coxswain Spurgeon said that he felt it likely that the *Agnes Cross* would have been needed and that delay would have occurred had the four of them not got round earlier. The Coastguard LSA shed was flooded and so the lifeboat would have been the only source of help had the need arisen. The Institution sent down 10s.0d. each for Spurgeon and the two Rose brothers but as the Mechanic was a full time paid employee he did not get extra.

In January 1926 George W Ayers, the Second Coxswain, retired. He had first been connected with the lifeboat in 1882 as a shore helper and was first in the crew for the service to the schooner **Marie** of King's Lynn on 18th November 1893. He had served for 45 years and the Institution awarded him a pension of £5.12s.6d per year together with a framed certificate of service. Henry G Rose of Nile House, Old Nelson Street was appointed in his place.

The records show that 1926 was a quiet year for the lifeboat with only five calls all of which involved the boat in standing by while others sorted out the casualties.

One of these incidents involved a small cutter from **HMS Godetia** (the fishery cruiser). On Saturday 3rd April the boat was seen in difficulties in a strong tide and was shipping water. The occupants (five sailors, a civilian man and a lady) signalled that they were exhausted and wanted to land. The coastguards advised that they would probably capsize in the breakers. Coxswain Spurgeon launched the *Agnes Cross* and reached them at 4.0pm to find that one of the railway tugs had taken the cutter in tow. The lifeboat stood by until they were safely in the harbour. The Honorary Secretary was indignant that the tug had rescued them and suggested to the Institution that a letter of protest be sent to the London & North Eastern Railway Co. When HQ sent down the payments they refused to send such a letter pointing out that the tug responded to people signalling for help and that such action should be commended.

On Monday 4th October 1926 the Ostend sailing trawler O.164 **Pattrick** was seen to go aground on the Newcome Sand. She was kept under observation but no lights or movements on board were seen and so, at 7.0pm after consultation with the Coastguard, the *Agnes Cross* was launched. When they came up with her the vessel was in darkness and seemed abandoned.

The lifeboat got alongside and eventually some of the trawler's crew came on deck and declined assistance with much abuse. Nevertheless Albert Spurgeon decided to stand by and an hour or so latter the vessel bumped off and sailed away. In his report for this service the Coxswain said that the searchlight (given by Miss Burrows – see page 1) had been used for the first time and had been of great assistance.

A busier year followed in 1927 with seven sailors having particular cause to be grateful. On Saturday 23rd July the Belgian trawler O.30 **Yolande** was seen by the Coastguards to be aground near the wreck of **HMS Spider** on the North beach and making signals of distress. The **Spider** had gone ashore in 1914 and had caused difficulties ever since. The *Agnes Cross* was launched and on arrival Coxswain Spurgeon manoeuvred her over the wreck of the **Spider** and alongside the trawler. He was preparing to take off the crew when the tug **Despatch** came up and he saw that there was a chance of saving the vessel. He took a hawser from the tug and this was made fast to the stern of the trawler which was rolling and shipping heavy seas. After an hour or so the tide had made sufficiently for her to be floated off. The tow was transferred to the bows with the help of four lifeboat crewmen who had been put aboard and although she was making water very fast the **Yolande** was successfully towed into the harbour. She was a new vessel and had been making for Lowestoft to land her catch.
Subsequently the share of salvage for the *Agnes Cross* was assessed at £75, giving each crew member £3.16s.8d with the 12 of them having to contribute to the 14s.3d sent to the Institution for the petrol used. The crew were not paid by the RNLI in salvage cases but were deemed to have borrowed the boat for the purpose of salvage.

The *Lowestoft Journal* of 5th November 1927 reported that Coxswain Spurgeon and nine of his crew were to appear on a lifeboat as part of the Lord Mayor of London's Parade. Fifty years later the *Lowestoft Journal* of 1st April 1977 gave the explanation: the Lord Mayor, Sir Charles Batho, had married Bessie Parker of Lowestoft and Sir Charles wanted the crew to "fly the flag for Lowestoft". Afterwards Sir Charles had commissioned the Mayor of Lowestoft, Mr Arthur Evans, to give the men "a good time". They had been entertained at the Suffolk Hotel and had thoroughly enjoyed themselves.

Monday 21st November 1927 dawned with leaden skies and strong winds which increased, until by dusk there was a whole easterly gale blowing with huge seas battering the coast. Any caught on this lee shore would have wished themselves elsewhere.

Albert Spurgeon was keeping watch from the lookout and at 4.0pm he saw a vessel battling against a strong tide and the storm. She missed the harbour entrance and went aground on the outer bank near the South Pier. While the *Agnes Cross* was launching, the casualty LT96 **Lily of Devon**, was being driven slowly towards the main bank where there was a concrete breakwater. Arriving at the wreck, Spurgeon dropped anchor and veered down in an attempt to get close. After four attempts the lifeboat grounded and was filled by the heavy seas. At this point we may take up the story from Albert Spurgeon's log-book:

> "The crew had to hang on to prevent themselves going overboard and I was knocked down but luckily escaped injury. The men on the smack were shouting and hanging on for dear life. The next sea lifted the lifeboat and, as it was now or never (we were within 30 yards of the concrete breakwater), I got alongside and grounded again. The sea knocked us onto the wreck with a terrific crash right under the port quarter breaking the rubbing strake, bending the ridge, damaging the fore part of the engine casing and smashing the starting gear. Two of the crew had a narrow escape from being killed. We hauled on the cable with the engine going full out, floated clear, got the three off the smack, lifted the anchor and landed three men."

A wonderful style of prose to describe what must have been a hair-raising experience for everybody. The Second Coxswain, John Rose and the Motor Mechanic, Ralph Scott were particularly commended by Albert Spurgeon for the way they carried out his orders.

The *Agnes Cross* was taken up to the yard of J W Brooke for repairs to the damage.

On 17th December 1927, Mr G F Shee, Secretary to the Institution, wrote to Mr Sydney Taylor the Lowestoft Honorary Secretary concerning the service to the **Lily of Devon**. He said, "In the opinion of the Committee of Management this rescue was due to the courage, skill and daring of the Coxswain and they have decided to mark their appreciation of his conduct by awarding him the Silver medal of the Institution with its 'Thanks on Vellum' and an additional award of £1.11s.6d to him and each of the 11 crew members with a special award of £1.10s.0d to the permanent Motor Mechanic."

After the report of this service in his log-book, Albert Spurgeon wrote the following:

> To Remember Always
> (after Lily of Devon)
>
> Life is a battle to lose or to win
> Are you a fighter or do you give in
> Running away when disaster seems near
> Making surrender to failure and fear.
>
> Go forth and conquer, You can if you will
> You have a destiny you must fulfil
> Don't be discouraged if progress seems slow
> You may be wounded but put up and show.
>
> Do not acknowledge despair and defeat
> Fight the gale forward and never retreat
> Rise up undoubted when heavy seas descend
> Fight for the right and you'll win in the end.
>
> (When danger nigh you'll all stand by
> Ready, aye, ready)

Albert wrote after this entry – "Wish you as my figurehead."

In the *Lowestoft Journal* of 31st March 1928 there was a report of the presentation of the Silver medal to Albert Spurgeon at the RNLI Annual Meeting in London:

"When he presented the medal HRH The Prince of Wales was heard to laugh heartily," and the report goes on, *"The News Chronicle* described it thus."

> 'Everyone in Central Hall wondered what Coxswain Spurgeon had said. The Prince had just pinned the medal on the breast of the smiling brown faced seaman and was about to shake hands when Coxswain Spurgeon suddenly said, "Will you come to Lowestoft, Sir?" "I should like very much." The Prince replied. Coxswain Spurgeon, not satisfied that The Prince really would visit said "We have some very nice Ladies in Lowestoft Sir, I am sure you would like them." The Prince burst into laughter and promised to do his best and still smiling, told the French Ambassador, who was close by, what the Lowestoft seaman had said.'

So ended another courageous episode in the annals of the Lowestoft Lifeboats, with Albert Spurgeon receiving the 19th RNLI Silver medal to come to the Lowestoft station.

The old **Lily of Devon** must have been tough, for 31 years later the *Lowestoft Journal* of 8th August 1958 reported seven muffled thumps and columns of water providing an unexpected thrill to hundreds of holidaymakers on the South beach. The charges were fired by Mr F Jensen recently returned home from oil exploration in the Persian Gulf. The Borough Engineer, Mr G A M Gentry, said that there had been a number of attempts to remove the wreck but that this was the first with explosives. The wreck was visible on only a few days each year but there had been complaints from people who had cut their feet on it.

In 2006 while this was being written, John Soanes, a friend of the author, was able to take photographs of some still remaining bits of the wreck during very low tides in the autumn. So parts of the vessel still survived after 79 years.

At the close of the year on Tuesday 27th December 1927 the *Agnes Cross* had an uncomfortable service to a dismasted vessel some 10 miles out. There was a whole gale blowing from ExN and Coxswain Spurgeon was keeping watch with the Coastguards. Around 2.0pm he saw through his telescope a sailing vessel in difficulties. A little while later he saw that a large flag had been hoisted in the rigging – a call for help. The *Agnes Cross* launched at 2.30pm into huge seas and only after a hard struggle did she come up with LT37 **Wide Awake**. The smack was in great peril with the mainmast gone and sails and rigging hanging over the lee side. The lifeboat came alongside but the crew of the smack refused to leave. At this time she was drifting rapidly towards the Holm Sand and so the Coxswain decided to stand by. Eventually the tug **Despatch** arrived and got a line aboard for a tow back to Lowestoft. Once they were underway the *Agnes Cross* followed and arrived back at 5.45pm with the crew thoroughly drenched having crossed the sands both out and back with the boat shipping heavy seas.

The remains of **HMS Spider** (wrecked in 1914) were to cause problems again early in 1928. After dark on Sunday 15th January a vessel went ashore on the North beach and the *Agnes Cross* was launched about 7.0pm in a moderate south-easterly gale. On arrival Coxswain Spurgeon found that the casualty, LT382 **Colinda**, was only a few yards from the **Spider** wreckage. The smack had bumped over the Outer Bank and it was only with great difficulty that the Coxswain was able to manoeuvre around the jagged remains and get alongside. The Skipper said that he had a crew member with a broken ankle sustained three days earlier and so the man was taken aboard the lifeboat and landed ashore for the hospital. The rest of the crew elected to remain with the smack. Later she knocked onto the Inner Bank and was leaking so badly that the crew burnt flares and the lifeboat attended again taking off the three remaining crew. Later, the **Colinda** was refloated and salvaged by the tug **Lowestoft**.

Walter Allerton, a lifeboat crew member, had fallen heavily while running round the dock after the maroons had gone and had, instead of going with the lifeboat, to attend hospital with a badly injured nose. For the same service the RNLI Committee of Management noted that Jack Rose, the bowman, had left his sick-bed to take his place in the crew and they sent him a Letter of Appreciation.

The *Lowestoft Journal* for 21st January carried a note that "The wreck of **HMS Spider** has stood out high and dry during recent exceptionally low tides together with the remains of **Corona**, **Irene**, **Diligence**, **Ranger**, LT167 **Belle** and **Scotia**." "A regular graveyard." the reporter noted. Later at a meeting of the Lowestoft Borough Council it was reported that prolonged easterly gales at Christmas had destroyed the Railway ballast groyne and placed a heavy strain on the North beach defences. The Borough Engineer felt that failure to remove the 'Spider' wreck had contributed to the loss of the groyne.

At midsummer a little incident confirms that some people had no more sense in the "old" days than now. At teatime on Friday 22nd June 1928 the Coastguard reported a small boat in trouble about five miles out. Some of the lifeboat crew had been watching her for a while and at 5.45pm the *Agnes Cross* launched and came up with the one ton yacht **Ugly** at about 6.30pm when she was in the vicinity of Scroby Sands having been driven by strong SW winds. The owner, a Mr Hailey of London, said he was on a cruising weekend and at first refused any help. However, a sudden squall nearly capsized him and he got into the lifeboat while his boat was taken in tow. The boat was only 15 feet long and had no oars. He was relying entirely on sail!!

Three days later on Monday 25th June the town was no doubt roused from its slumbers when the maroons were fired at 3.45am. A small two ton fishing boat LT683 **Mica** owned by the Second Coxswain of the lifeboat was on fire about eight miles south of Lowestoft. The *Agnes Cross* attended and found that a crew member of **Mica** had been filling the petrol tank when some of it splashed onto a hurricane lamp with the inevitable consequences. However the burning items were thrown overboard and it was petrol burning on the surface of the sea that attracted the attention of watchers on shore. There was little damage to the boat and the lifeboat returned to port.

Later in the summer the *Lowestoft Journal* of 11th August 1928 reported that the Lifeboat Flag Day had been held the previous Saturday. "It was an altogether miserable, wet and windy day but the sellers stuck to their posts." It seems that, just as to-day, the lifeboat gave a display for the benefit of people on the South beach and apparently despite the conditions it attracted a large crowd. In total the collection raised £79.3s.9d, with the men from the Lord Kitchener home collecting £30.4s.8d. of this.

A lengthy service on the night of 26th/27th August 1928 resulted in the saving of a crew of eight. Late in the evening of Sunday the 26th a vessel was heard making urgent signals for assistance on her siren. The *Agnes Cross* launched at about 10.30pm and found that the steamship **E Rose** of Great Yarmouth had grounded on the Outer Bank with heavy seas breaking over her at times. Her Master asked Coxswain Spurgeon if he would stand by and in the early hours of the following day she knocked off onto the Inner Bank in the, by then, increased winds. A tug, which was also at the scene, was unable to help because of the shallow water. Knowing that the casualty was very close to the remains of **HMS Spider**, Spurgeon took the lifeboat between the submerged remains and the shore and made fast alongside the steamer. At 4.30am the **E Rose** was thrown onto the jagged remains and holed in the engine room. The entire crew of eight jumped into the lifeboat as the stern of their ship went down. Lines to her were hastily chopped and the *Agnes Cross* pulled clear and made for the harbour to land the shipwrecked mariners. The *Lowestoft Journal* of 15th September 1928 reported that the 106 ton vessel was salvaged by the Grimsby firm of Charles and Co., who sent the salvage vessel **Osprey** under Captain Banks. The hole was patched by divers, some cargo thrown out, the vessel refloated and towed into the Waveney Dock for temporary repairs. She was later towed to the River Humber for permanent repairs.

Close to the end of the year a service by the Coastguard Lifesaving Apparatus team, in which the lifeboat stood by, resulted in several letters of thanks and commendation for the *Agnes Cross* and her crew.

At dusk on Sunday 30th December 1928 Coxswain Spurgeon saw a steam trawler crossing the Inner Shoal in a whole NE gale with very heavy seas running. The vessel, IJM.80 **Elnet** of Ijmuiden, was struck by a heavy broadside sea and she hit the Outer Bank about 100 yards south of the lookout. She immediately sounded her whistle for help and three maroons were fired for the lifeboat and the Coastguard LSA. The *Agnes Cross* launched and came up with the trawler which was being swept by heavy seas and was by now in shallow water. Coxswain Spurgeon let go his anchor and attempted to veer down on the **Elnet** but after three attempts the lifeboat was hit by two broadside seas and filled to the gunwales. The lifeboat crew hung on for dear life and, after checking they were all safe, Spurgeon had another go only to be washed away again, hitting the bottom in the process. The trawler was by now in even more shallow water and the *Agnes Cross* could only stand off and illuminate the scene with her searchlight. The crew of eleven could be seen crouched in the bows of the **Elnet.** Eventually the LSA crew managed to get a line on board her and were able to rescue all the men. Coxswain Spurgeon reported that both the lifeboat and her crew behaved splendidly and he gave special thanks to the Motor Mechanic and his new assistant. What a baptism it must have been for him – howling wind, darkness, heavy seas, shallow water and rain squalls!

The lifeboat crew received thanks from the Board of Trade for standing by and using the searchlight to help the rescuers. A letter was received from the Consul-General of The Netherlands and one of sincere thanks from The Netherlands Legation in London to The President of the Board of Trade, the Rt.Hon. Sir Austen Chamberlain KC, was copied to the RNLI in London who forwarded it to Lowestoft. The Lowestoft Coastguard Lifesaving Apparatus team were awarded the Challenge Shield for the best service of 1928. The thanks of HM Coastguard at Great Yarmouth were received by the Lowestoft lifeboat together with an invitation to the presentation ceremony.

An interesting sequel appeared in the *Lowestoft Journal* of 25th March 1988 in a letter from Coastguard Sector Officer Mr D K Gurton. He was responding to an earlier note about the **Elnet** and said that he still had the brass plaque recording the award of the shield to Lowestoft in his office. He noted that the rescuers worked standing in the sea because the press of spectators on the beach prevented the gear from being laid out in the usual way. So, even on a terrible December night in 1928, the sound of the maroons still brought people out of their warm homes to witness some outstanding bravery. Sadly nowadays, with the advent of pagers, most of us never know when our lifeboat goes out until perhaps, we read it in the papers.

The following year, 1929, was a quieter one with the lifeboat standing by a number of times. The year's tally of people saved, three, were all from the same small dinghy.

On Friday 18th April three small boys in a 12 foot dinghy, the **Boy Fred**, set out from the harbour after lunch. By 3.0pm the Coastguard saw that they were on the inner edge of the Newcome Sand with a sail hoisted. By 4.0pm they had got as far as the East Newcome buoy - 2¾ miles out. They were by now in danger in the strong WSW breeze and the *Agnes Cross* was launched. When she came up with the boat the boys were described as "distressed", with only one oar, a split sail and no anchor. They were taken on board the lifeboat and their boat taken in tow for the harbour. The *Lowestoft Journal* of 20th April reported that, "...once in the harbour one of the lads calmly sculled the boat away."

Also in April, Mr Ralph Scott, who had been Motor Mechanic since the *Agnes Cross* first came to Lowestoft, moved to Holyhead. He became Mechanic of their new 51 foot Barnett type lifeboat *AED* (ON717). He had been awarded the RNLI Silver medal for the **Hopelyn** rescue in 1922.

The *Lowestoft Journal* of 10th August 1929 reported that the Lowestoft station had been awarded its Centenary Vellum. It was presented to the Town Clerk, Mr C Ashton-Stray, by Sir Gervais Rentoul the MP for Lowestoft who was also a member of the RNLI Committee of Management. Sir Gervais said that it was a privilege to present the Vellum for 100 years, although actually it was considerably over 100 years. In fact, as readers of Part 1 of this book will know, the first lifeboat was established at Lowestoft 128 years earlier. During research for this volume, the 200th Anniversary Vellum was presented to the Lowestoft station in 2001 at a special ceremony at the Marina Theatre in the town.

The remains of **HMS Spider** continued to cause added problems for vessels running aground on the North beach. On Saturday 4th October 1929 two sailing trawlers, LT1223 **Shrublands** and LT1097 **Kestrel** grounded close to the remains of the **Spider**. The **Kestrel** drifted off and, narrowly missing the **Shrublands,** hit the jagged remains. The *Agnes Cross* launched at 1.55pm and, on arrival, dropped her anchor and veered down to the trawler which was right in the centre of the 'Spider' remains. However, the Skipper asked the Coxswain to stand by as she was filling. By late afternoon the Owners and others boarded her and pumping out commenced with a team of over 30 people on board. Eventually the **Kestrel** slipped off the wreckage of the **Spider** and grounded again where she remained until, at 8.0pm, a tug got a line on board and towed her to the harbour where she sank. The **Shrublands** had been towed afloat earlier. The lifeboat returned to her moorings at 8.40pm.

During November 1929 the AGM of the Lowestoft Branch of the RNLI was held. It was noted that the firing of maroons and shore signalling duties had been taken over by the Coastguard from 1st October. Consequently the station signaller was no longer needed. John L Burwood had been appointed signaller in 1906 and had a total of 42 years service with the lifeboat: 18 in the crew and 24 as signaller. He was presented with a framed certificate and was awarded an annual pension of £3.11s.3d.

We have seen numerous times how the remains of **HMS Spider** caused difficulties to vessels running ashore and to the rescuers. In November 1929, Sir Gervais Rentoul MP, asked the President of the Board of Trade about the trouble caused, pointing out, "…that nine vessels have run into it so far." The Board of Trade replied that the Local Authority had power to remove it if they wished. Perhaps this question in the House of Commons was prompted by the MP's earlier visit to the town.

1.2 - The 1930's

During 1930 calls for the lifeboat were down to five with seven people being saved. On Wednesday 7th May the four ton fishing lugger LT104 **Enterprise** went on the outer edge of the Barnard Sand with her gear fast on the bottom and the engine broken down. The *Agnes Cross* launched just after 4.0pm and found the two crewmen exhausted by their efforts in the strong NNE wind and heavy breaking seas. Two lifeboatmen were put aboard and made a towing hawser fast. The fishing gear was recovered and the casualty towed into the harbour by 6.10pm. The RNLI HQ disputed the need for the launch and only paid the exercise rate – 12s.6d. per man.

Later the same month, on the 25th, the Coastguards assisted to safety three small boys who got into difficulties in a rowing boat, the **Tom Bowling**, off Ness Point. After the boys had been safely got to the beach, three men launched the rowing boat off the beach intending to take it to the safety of the harbour. Almost immediately they got into trouble. The *Agnes Cross* had been standing by while the boys were helped and she went to give assistance, taking the boat and three men to the harbour.

Earlier in the year ex-Coxswain John Swan had attended the BBC in London to make a broadcast appeal on behalf of the RNLI. This was transmitted on Sunday 2nd March 1930 and by May 22nd had raised £724. Since radio (the wireless) was then so very new, I wonder if this was the first such appeal for the RNLI. I have a receipt which John Swan signed for his expenses - £1 rail fare to London and two nights in the Capital at 10s.0d per night.

The *Norfolk News and Weekly Press* of 29th March reported that a portrait of John Swan was unveiled, at a meeting of the Loyal Lacon Lodge of Oddfellows at St. Margaret's Institute, to commemorate his 50 years of membership. The following week the same newspaper reported that he had been awarded the RNLI Gold Badge for his voluntary services in raising funds for the RNLI.

Later in the year the *Lowestoft Journal* of 25th October 1930 reported that Mr W N Rose had received a framed certificate for 52 years service with the lifeboat crew. He had first sailed in the lifeboat **Samuel Plimsoll** (ON22) in 1877 to the brig **Hope** and a month later to the lugger LT132 **Pet** when he took part in the rescue of the crew of 11 and a dog. Also during 1930 it was reported that William Cooper who had joined the lifeboat crew in 1863, aged 17, under Coxswain Robert Hook, also applied for a certificate. At the end of 1930 Mr H T Burgess, a young long-shore fisherman was appointed to the lifeboat crew. This was Harry Burgess who appears later in this book.

On Monday 24th November 1930 there was a whole SE gale blowing with heavy seas making the approach to Lowestoft harbour very tricky. Just before noon the steam drifter LT1118 **Impregnable** was crossing the bar when she was hit by a huge sea which broke the steering gear and pushed her alongside the North extension. She was kept off the stonework by the backwash from the big waves hitting it and the Skipper was able to go astern and get into whole water where he was able to carry out temporary repairs before getting in safely. The lifeboat crew were mustered but the boat did not launch. Shortly afterwards the smack LT1097 **Kestrel** was in trouble three miles out with a broken gaff and torn mainsail. She ran before the gale with a reefed mizzen, jib and foresail into the channel between the Holm and Newcome Sands. The lifeboat crew remained assembled until the tug **Despatch** towed the **Kestrel** safely in by 3.30pm.

While the **Kestrel** had been kept under observation so, also, had a vessel making in from the north. By late afternoon she had got into the heavy seas running on the Inner Shoal, where a particularly nasty one broke aboard and carried away the stay foresail. Distress signals were hoisted. A Dutch Fishery Cruiser was about a mile off but apparently did not see his fellow countryman (the casualty was the fishing lugger VL.88 **Adriana I**). It was thought later that the Captain was having too many problems looking after his own vessel in the awful weather.
By 4.30pm the **Agnes Cross** was away and on getting alongside the Dutchman the Skipper reported that he had engine trouble, had lost all his canvas and needed assistance. Eventually a tug arrived and the lifeboat stood by until the **Adriana I** and her crew of 17 were safely in the harbour. The lifeboat had an extra crew member for this service. Mr W Spinks had fallen between the quay and the lifeboat and was dragged aboard as she was pulling away without Albert Spurgeon being aware that he had an extra man. Mr Spinks carried on soaking wet and, to add insult to injury, the payment list for this service suggests that he may not have got any money for his trouble!

During 1931 there was little lifeboat activity, with only one service which resulted in the saving of two people and a boat. On Friday 18th December the fishing boat LT104 **Boy Reggie** suffered engine failure near the Holm Sand. The *Agnes Cross* launched, without a full crew, at about 2.30pm and towed the boat, which was making water, into the harbour and beached her on the hard.

The *Norfolk News and Weekly Press* of 23rd May 1931 reported that Messrs Hobson and Co., auctioned two lifeboats on the North beach. The former Winterton reserve boat *Eleanor Brown* (ON589) was withdrawn from sale, whilst the former Cromer No.2 boat *Louisa Heartwell* (ON495) was sold to a Mr Green for £55. Various boat fittings were also sold for a total of £6.13s.6d. Later, a bid of £50 was accepted from Mr Sam Long of Blakeney for the *Eleanor Brown*. Three oars, two foresheets and various lengths of rope were provided by the Lowestoft Branch of the RNLI to replace those missing from the boat. The new owner sailed her away on 30th May. She had been built by Thames Ironworks in 1909 and was $44\frac{1}{2}$ feet by $12\frac{1}{2}$ feet.

It appears that both boats had been stored, temporarily, in the old lifeboat sheds on the North beach. In July these sheds were cleared of gear and rubbish prior to their being sold. Albert Spurgeon had paid 12s.6d for two 30 foot launching spars and a quantity of firewood. While making plans to sell the sheds the RNLI had been in correspondence with Messrs Foster, Calvert and Marriott of Norwich, Solicitors to the Lord of the Manor of Lowestoft. On 26th April 1839 there had been a conveyance of land from the then Lord of the Manor to Sir Thomas Gooch and others who were Trustees of the Suffolk Humane Society. It was a piece of waste land to be used for "the safe custody and preservation of the Lifeboat and her stores, tackles and appurtenances." There was an annual rent of two shillings, payable in perpetuity to the Lord of the Manor. Upon this land, eventually, the RNLI had built two sheds, one in 1858 which was 56 feet long and the other in 1870 being 41 feet long. They were sold and much later, in 1972, were demolished in the Lowestoft Beach village clearances.

During the autumn, after many years of causing havoc on the North beach, a start was made on clearing the remains of **HMS Spider**. The *Norfolk News and Weekly Press* of 12th September reported that dynamite would be used.

In October the *Lowestoft Journal* of the 17th reported that the former bowman, Mr Jack Rose, had received a Certificate of Service for 53 years service in the lifeboat. He had the RNLI Bronze medal for the **Hopelyn** rescue in 1922. He had joined the crew in 1878 and had been involved in the rescue from the barque **Berthon** in 1882 after the unfortunate events of "Black Saturday" which were recorded in Part 2 of this story.

During the morning of Wednesday 13th January 1932 a severe south-easterly gale was blowing creating very heavy seas and causing consternation on board vessels in the Roads. Just after 10.0am the Coastguard at Gorleston telephoned Coxswain Spurgeon to say that a barge anchored off Corton was signalling for help. The *Agnes Cross* launched at 10.30am and went north towards the barge. She was the **Servic** of London being swept by heavy breakers and dragging towards the shore. After a while it appeared that the anchors found better ground and the Skipper asked the lifeboat to stand by until a tug arrived to take her to Yarmouth. After the tow was underway Albert Spurgeon made the rounds of other anchored vessels and finding them all safe he returned to Lowestoft after a short but rough passage.

During March 1932, John Ward, the last Coxswain of the Pakefield station died. The *Lowestoft Journal* of 26th March recorded that he had been in the crew for 30 years before retiring in 1922 when the station was closed. He had been presented with a Certificate of Service at a ceremony at the Hippodrome cinema in Lowestoft. At that time the slipway at Pakefield had been washed away and as a motor lifeboat was to be allocated to Lowestoft it was decided that the Pakefield station was not needed and it was closed after 82 years. John Ward was a well known local boatman and had worked the sailing pleasure boat **Britannia** from Lowestoft and had been much in demand as a pilot for large yachts especially during regatta weeks. He had been appointed Coxswain in 1910.

Even in better weather problems at sea are never far away. On Sunday 12th June 1932 about two miles SE of Southwold the barge **Scone** of Rochester caught fire. At about 11.0am Albert Spurgeon received a message to say that the crew were fighting it as best they could. In view of the distance, about 14 miles, he set off immediately and came up with the barge at 12.40pm by which time the fire was out. After being assured that everything was in order the lifeboat set off for home, while the **Scone** continued her voyage from Hull to Gravesend. Southwold lifeboat *Mary Scott* (ON691) was off station for overhaul and so the Lowestoft boat was covering for her.

A few weeks earlier the *Lowestoft Journal* of 14th May reported that John Swan, ex-Coxswain of the Lowestoft boat, was to take charge of an exhibition of models of the latest types of lifeboats. This was to be held in the showrooms of Citroen Cars Ltd in Piccadilly, London by courtesy of M.Andre Citroen who was a Life-Governor of the RNLI. At the end of the 20th century the big motoring connection with the RNLI was Volvo Cars Ltd. The present Gorleston lifeboat **Samarbeta** (ON1208) was funded by Volvo Cars Ltd, her name being Swedish for "working together".

To return to the Lowestoft story. On Friday 28th October 1932 there was a strong gale blowing from the NNE with very heavy seas crashing onto the beach. During the afternoon Coxswain Spurgeon was informed that a small 'object', believed to be a boat, was in difficulties south of the Coastguard lookout. By now the wind was force 8 or 9 and after getting word that a small boat, LT1214 **Sonny Boy**, was missing Albert Spurgeon decided to launch. The *Agnes Cross* got away about 3.0pm and found **Sonny Boy** sheltering on the inside of the Newcome Sand well filled with water after shipping some seas. Her crew of three were taken onto the *Agnes Cross* and their boat was taken in tow for the harbour with all being safely back by 4.30pm. The *Lowestoft Journal* of 5th November recorded that the small fishing boat was after sprats and had been crewed by George Nobbs of Seago Street, Thompson Swan of May Road and Tom Moore of Hill Road.

A service on Tuesday 1st November to the wooden drifter K27 **Bezaleel** completed the work for 1932. She had sailed just after midnight and shortly afterwards her engine failed. She drifted onto the Outer Bank and burnt flares for assistance. When the *Agnes Cross* reached her she was drifting towards Ness Point. Just as her crew were going to get on board the lifeboat their Engineer managed to restart the motor and she was able to resume her voyage in a strong WNW wind.

In early 1933 the *Lowestoft Journal* of 25th February reported the retirement of the Harbour Master Captain J G Gwilliam. He had followed his father into the Great Eastern Railway Co but later left and qualified to command square-rigged sailing ships. He subsequently returned to the GER on their ferries running from Harwich to Antwerp and The Hook of Holland. He was succeeded by Captain L A Rhodes. A presentation was made to Captain Gwilliam by the Berthing Master George Munnings.

Just after lunchtime on Friday 31st March 1933 a boat was seen trying to enter the harbour but having drifted past the entrance she anchored. The pilot boat attempted to tow her but could not and she returned to harbour and informed the Coxswain that help was needed. The *Agnes Cross* was about to leave on a routine exercise and so help was offered and accepted. Two of the lifeboat crew were put aboard the disabled motor cruiser **Francomme** and a towing hawser made fast. The boat was safely towed to the harbour and the lifeboat continued with her exercise. The RNLI records show that the crew were paid 7s.6d. each for the exercise with no mention being made of the **Francomme**.

During May and June 1933 the *Agnes Cross* was off station for her annual overhaul at the yard of Messrs JW Brooke. The relief boat *City of Bradford* (ON680) had a call on Thursday 8th June. In the late morning, workmen for the Lowestoft Borough Council were carting shingle on the North beach in dense fog, when they saw a ship heading for the shore. They shouted warnings, which luckily were heard on board and the ship went astern and dropped anchor. By now she was lost to sight and a message was taken to the Coastguards and the lifeboat launched just after 11.0am. She found the vessel, the ss **Effra** of London, had struck the sand near Ness Point but had got off and was able to resume her voyage after lying at anchor for a while.

A bizarre incident happened later, after the **Agnes Cross** had returned from refit. On Wednesday 21st June information came to Albert Spurgeon that two canoes were in difficulties off Kessingland. He launched promptly, since there were heavy seas driven by a moderate southerly wind. On arrival they found one canoe stove in with its occupant hanging on to the other canoe which was crewed by a German girl and a small dog. All three and the canoes were taken on board the *Agnes Cross* and safely taken to Lowestoft. The *Lowestoft Journal* of 24th June reported that the girl was Fraulein Freda Mayel who, together with her companion Herr Edward Engert, were trying to circumnavigate the British Isles in collapsible canoes. They had reached Southwold on Tuesday and set out again on Wednesday but off Kessingland Herr Engert's canoe capsized and he was struggling in the water while the girl recovered his canoe. He managed to hang onto her canoe but she was unable to make progress with the extra load and signalled to shore for help.

On the following day the couple gave a talk about their experiences at the Grand cinema in Lowestoft. Frau Mayer left on the Friday to continue her voyage, escorted for a while by Coxswain Spurgeon in his own boat but without Herr Engert who decided not to continue. It would be interesting to know if she completed this epic.

I am reminded of Major Ewan Southby-Tailyour who, while serving with the Royal Marines in the Falkland Islands in 1978, had used his leisure time to sail extensively round the islands mapping every major bay and inlet on the long coastline. His notebooks were to prove invaluable during the Falklands war of 1982. Who knows what Frau Mayer had been doing all those years earlier!

On Thursday 7th December 1933 a gale from the ENE had sprung up before breakfast time. At 6.0am the Coastguard reported a sailing trawler ashore at the North Pier extension with the seas breaking over her and the crew shouting for help. The *Agnes Cross* launched and succeeded in getting alongside LT910 **W E H** but the heavy seas knocked her away and at the same time knocked the casualty off the sand and she drifted into the harbour leaking badly. The lifeboat remained until she was safe. Shortly after, at about 7.0am, word was had that a small fishing vessel was in danger about four miles away. This time the lifeboat found LT379 **Marjorie** on her way in and shipping heavy seas. Coxswain Spurgeon put *Agnes Cross* on the weather side of her and escorted her across the Newcome Sand and the harbour bar to safety.

At the end of December, the difficult harbour entrance was improved somewhat by dredging carried out by two London and North Eastern Railway Co dredgers **Pioneer** and **Mudsucker**.

It turned out that 1934 was a very quiet year with only two calls both of which involved standing by.

In the early hours of Sunday 1st July the ss **Marion Traber** ran aground a mile north of the Coastguard lookout. At around 4.0am the **Corton** lightvessel began firing her guns to alert those on shore to the casualty. The *Agnes Cross* was not officially on service following her refit at Fletcher's yard in Oulton Broad. However, after discussions with the Hon. Secretary, it was agreed that Coxswain Spurgeon should launch. On arrival at the ship it was found that help was not needed but the lifeboat stood by while the vessel's small boat laid out an anchor to help pull her clear. When she was refloated the lifeboat guided her clear of the banks.

This service led to correspondence with RNLI HQ who firmly pointed out that "not on service" means just that and further that the Gorleston lifeboat had launched to signals from the lightship. It was also pointed out that the casualty had not made signals of distress and that Coxswain Spurgeon had ignored signals of recall made from the shore. In the end it was agreed to treat the launch as an exercise and each man got the summer rate of 5s.0d. It is perhaps worth noting that the country was in the grip of an economic depression at this time and no doubt several of the crew may have had no permanent job. Consequently, Albert Spurgeon – a compassionate man – might have seen a marginal case as the opportunity of a few extra shillings for his men.

Earlier, the *Lowestoft Journal* of 9th June 1934 reported that delegates to the National Savings Assembly, held at the Royal Hotel, had attended the South Pier at the conclusion of their meeting. Their Chairman, Lord Mottistone, addressed the lifeboat crew members. Also present were: The Mayor (Cllr. W Mobbs), the Town Clerk (Mr C Ashton-Stray), Mr F Spashett (Chairman of the Lifeboat Committee), Mr Sydney Taylor (Lifeboat Secretary) and Lord and Lady Somerleyton. Lord Mottistone (Major-General JEB Seeley) shook hands with the crew and drew laughter when he referred to one of the most famous lifesaving crews meeting with one of the most famous assemblies of National Savers. A collection was taken which raised £8.6s.6d. Lord Mottistone was at that time a Vice-President of the RNLI, a member of the RNLI Committee of Management and Coxswain of the Brooke (Isle of Wight) lifeboat.

A much busier year followed in 1935 with 15 people being rescued. In the early part of the year the town was saddened by the death of former Coxswain John Swan, the holder of an RNLI Gold medal and two Silver. He had served for over 50 years in the crew and retired in 1924 thereafter devoting himself to fund-raising and publicity for the RNLI. He died on 20th February 1935 and was laid to rest beside his wife in St. Margaret's Churchyard with members of the ***Agnes Cross*** crew acting as bearers.

The inscription on his headstone refers to "John Thompson Sterry Swan VC,OBE ….." I suspect that "VC" is the Lifeboatman's VC – the Gold medal. His funeral and obituary were fully reported in the *Lowestoft Journal* for 23rd February and 2nd March 1935

Before that, on Wednesday 30th January 1935 the large Greek steamer ss **Pagasitikos** of Andros, bound from Alexandria for Hull with cotton seed, skins and bones ran hard and fast on the Newcome Sand in dense fog. The *Agnes Cross* launched at 9.25am and located the vessel lying with a heavy list to starboard. The Captain refused help but his ship settled more and more as the tide fell. Eventually four tugs **Burton**, **Despatch**, **George Jewson** and **Tactful** were on the scene but were not engaged. Later a German tug from Harwich and two other foreign tugs did succeed in refloating the ship, which was of 3457 gross tons, on the next day. The lifeboat had stayed until 7.0pm when having told the Captain that he would return if the weather worsened Albert Spurgeon took the lifeboat home.

On Tuesday 14th May, the relief boat *City of Bradford* was called to help a small fishing boat LT259 **Joan** when she shipped a sea which swamped the engine. There was a north-easterly gale blowing. The two crew of the fishing boat were Robert Thurston aged 80 years and Walter Allerton aged 75. The *Lowestoft Journal* of 18th May reported that 'both veterans declined to leave their boat which was pumped out and towed in by the lifeboat'.

In the autumn a very smart service was carried out to the sailing smack LT975 **Challenger**. In the early evening of Sunday 20th October she ran onto the bank at the end of the North extension and the *Agnes Cross* launched at 6.38pm in response to her flares. The Coastguard Lifesaving Apparatus team also attended and got a line aboard her. However, the lifeboat ran quickly in, got a tow rope fixed and was able to pull her off aided by the swell and towed her into the harbour. At the same time red lights were seen to the south and so the lifeboat set off but these proved to be lights on repair work at the foot of the cliffs.

During the foggy morning of Sunday 3rd November 1935 the small fishing boat LT83 **Golden Miller** was fishing for sprats when her engine broke down. The Coastguard informed Coxswain Spurgeon who went to the North Pier and through his telescope was eventually able to make out the boat with a flag tied to an oar and two men waving. The *Agnes Cross* got to her about noon and soon had her safely in the harbour.

At the end of the year a strong SE gale caused problems for the steam trawler LT153 **Rochester**. At 4.0am on Monday 30th December 1935 she was seen burning flares having run onto the Inner Shoal. Albert Spurgeon got the *Agnes Cross* away promptly but could see nothing of the casualty since all her lights had gone out under the force of the impact when she hit the shoal. The lifeboat searched for over four hours using the searchlight in heavy rain but to no avail and returned to harbour. It was later learnt that the Skipper had been taken ill and the Mate, in order to save time, had decided to cross the shoal. Soon after she had showed flares a big sea had knocked her off into deep water and the Mate wisely decided to lay off until conditions improved. The vessel was too far out for the lifeboat to find her.

The *Lowestoft Journal* of 4th January 1936 reported that lack of illumination played a second part in this service. The lights had fused at the lifeboat jetty and while getting on board, Charles Rose, son of the Second Coxswain fell between the boat and the jetty. He was dazed and could not cry out but hung on to the lifelines until his plight was discovered. The paper recorded that he cheerfully carried on despite being soaking wet.

In the RNLI journal *The Lifeboat* for October 1935 there was an interesting tale. At the end of January Albert Spurgeon and his crew travelled to Southend to bring back the relief boat **City of Bradford**. The journey home was a rough one with the boat continually swept by freezing seas on top of snow and hail. Mr Frank Carr, Assistant Librarian at the House of Lords, was a passenger on the boat and wrote of Coxswain Spurgeon, "He kept the helm most of the way and was magnificent. His cheery leadership in depressing conditions made one realise what a tower of strength he would be in arduous service conditions. He had an inexhaustible fund of stories and his hearty laughter shook the boat more than the seas."

The new-year, 1936, saw several calls which did not result in any service. However, one which did was to the former Wells-next-the-Sea sailing lifeboat **Baltic** (ON665). She had been replaced by a new motor lifeboat **Royal Silver Jubilee 1910-1935** (ON780) and had left Wells for survey and repair at Oulton Broad. She had been expected at Lowestoft during the afternoon of Saturday 22nd February. When she had not arrived by the early hours of Sunday permission was given for the **Agnes Cross** to launch. They found the Wells boat 16 miles out severely hampered by head winds and squalls and towed her to Lowestoft. The *Lowestoft Journal* of 29th February reported that Coxswain Neilson of Wells and his crew did not disguise their appreciation of Coxswain Spurgeon's thought which had saved them at least another seven hours hard work.

During the week before this service, the London & North Eastern Railway Co's tug **Despatch** made her last voyage from the port when she was towed away to the breakers after 61 years service. Since the advent of steam powered tugs at Lowestoft in the mid-19th century they had worked closely with the lifeboats towing them out to scenes of wrecks, sometimes in appalling weather and saving the lifeboatmen hours of hard rowing and sailing in the process. The railway companies made no charge for this service and indeed, at times of bad weather, the tugs were often maintained with steam up just in case of need by the lifeboat. The **Despatch** was being towed to the river Tees by the tug **Barton** but north of Flamborough Head the tow parted and could not be remade in the bad weather. The **Barton** continued alone. Apparently the derelict **Despatch** was picked up by a Latvian ship the ss **Velta** and towed to the Tyne. Also in the same week the railway steam paddle tug **Imperial** went to the breakers after 57 years of service.

On Sunday 17th May 1936 the steel sailing barge **Will Everard** of London got into trouble near the Scroby Sands. As the Gorleston lifeboat was off service the Lowestoft relief boat *J B Proudfoot* launched at 11.40am. She found the barge under tow by a tug and stood by until they were both safely inside Gorleston piers. This service was tinged with sadness. William Charles Butcher aged 63 years, a long serving crew member, collapsed and died while running round Hamilton Dock to take his place in the lifeboat. It was not realised before the boat went out and his comrades only found out when they returned. The *Lowestoft Journal* of 23rd May reported that his RNLI Bronze medal, for the **Hopelyn** service in 1922, was buried with him.

On the evening of Monday 3rd August three men and a boy got into difficulties in a rowing boat about a mile out being unable to make headway against the tide. The *Agnes Cross* got alongside about 8.0pm and found one man lying in the bottom of the boat and all four exhausted. They were taken into the lifeboat and their boat baled out before being towed back to harbour. The *Lowestoft Journal* of 8th August reported that passengers on the **Queen of Kent** returning from a day trip had a grandstand view of the rescue and cheered the rescuers.

On Sunday 4th October 1936 two youths, who had cycled over from Yarmouth, decided to row back home in a hydroplane hull they had earlier bought. They put their two bikes on board (!) but only managed a short distance outside the piers before they capsized. One of them, a boy named

Clayton, got ashore but the other was lost. The *Agnes Cross* launched at 4.10pm and searched for some time but only found two oars, a box and some gratings.

At this time in the 1930's the use of radio was gathering pace in all sorts of applications. The *Lowestoft Journal* of 25th July 1936 reported that wireless apparatus, which had already been successfully used in coal mines, might be used in lifeboats. It had been developed by Messrs LW and CE Hermes, a father and son. Trials at sea involved the *Agnes Cross* and the Aldeburgh lifeboat *Abdy Beauclerk* (ON751) and lasted about seven hours. Both inventors expressed themselves satisfied with the results and felt that their system could be the means of saving lives at sea.

Much later the *Lowestoft Journal* of 5th February 1938 reported that wireless in lifeboats would only be fitted in cabin boats – so there would be no installation in the Lowestoft boat.

During 1937 the lifeboat was only called in the months of May, November and December, during what was a quiet year.

During February 1937 Albert Spurgeon's Mother, Elizabeth, died at the age of 94. Her husband William, who died in 1905, had also been in the lifeboat crew. Mrs Spurgeon who lived at Russell Villas, Newcome Road in the beach village, had spent her life making fishing nets. The *Lowestoft Journal* of 27th February 1937 reported that, "…..she was rarely seen without the tools of her hobby." (!)

Twice in one week the lifeboat was called to a motor boat from the survey ship **HMS Fitzroy**. On Monday 10th May the launch, with one officer and seven men, struck submerged piling 200 yards off Ness Point and the engine was flooded. Coxswain Albert Spurgeon was at the lifeboat organising its decoration for the forthcoming Coronation, so the relief boat *City of Bradford* was soon away. The Coxswain lashed the lifeboat alongside the launch, valuable gear was transferred for safety and, with the lifeboat crew helping to bale, the launch was taken back to the survey vessel's berth in the harbour.

The Coronation of King George VI and Queen Elizabeth was held on 12th May and the *Lowestoft Journal* reported that the decorated lifeboat made a proud picture for the celebrations.

On Monday 17th May another launch from **HMS Fitzroy** got into trouble, this time with her propeller fouled, about five miles out. By the time the lifeboat got to her she was under tow by a motor herring boat. The *City of Bradford* stood by until they were both safely in the harbour.

On Monday 1st November 1937 the longshore boat LT379 **Terry**, with a heavy load of sprats, found herself in danger of being swept ashore in the heavy seas. Watchers at Pakefield saw her plight and were quick to call the lifeboat. The *Agnes Cross* launched at about 4.15pm and by the time they got to her the fishing boat was within 20 yards of the breakers and dragging her anchor. The Lifesaving Apparatus team were on the beach but the lifeboat got a line to her and was able to drag her clear. Buckets and pumps cleared the vessel and the *Agnes Cross* towed her Lowestoft.

In thick fog on Wednesday 1st December the Scottish drifter KY48 **Pilot Star** ran aground in the Stanford Channel. She knocked off later but lost her propeller. She made distress signals on her siren. The Coxswain launched the *Agnes Cross* at 1.35pm and headed for the Holm Sand but came upon the drifter half-way across the Stanford Channel. The drifter had hoisted a sail and with the lifeboat as escort managed to sail out to the NE Corton buoy. Here the Yarmouth tug **George Jewson** took her in tow under escort of the Gorleston lifeboat. The tug had actually gone out in respose to signal guns from both the **Corton** and **St. Nicholas** lightvessels.

Also during the year Henry George Rose of Old Nelson Street, the Second Coxswain, retired after 46 years service. He was awarded a pension of £4.8s.2d. per year. Arthur J H Swan was appointed in his place having joined the lifeboat crew in 1898 at the age of 17. At the end of the year Mr F W Cook the Motor Mechanic was promoted to the Southend station his place being taken by Mr G W Hitter, a Lowestoft man, promoted from The Mumbles station in Wales.

On Monday 14th February 1938 the *Agnes Cross* was called to the aid of the sailing smack LT1099 **Northern Queen**. The alarm had been raised by the Southwold station who were unable to launch because the harbour mouth was blocked by a steam pile driver which had been blown over by the NE gale. Albert Spurgeon launched the lifeboat at 8.45am and when she crossed the Inner Shoal she was swept by several seas, soaking the crew and damaging the engine casing. They found the casualty, with all her sails blown away, about 19 miles from Lowestoft with the Aldeburgh lifeboat standing by. Both lifeboats stayed until a tug arrived to take the **Northern Queen** in tow when they both escorted the tow to Lowestoft, the Aldeburgh lifeboat being unable to return to her station because of the severity of the gale.

Only four days later, on the 18th, the *Agnes Cross* was in action again. The smack LT961 **Lucky Hit** ran aground on the Newcome Sand after her engine failed. The lifeboat and the tug **Lowestoft** went out by which time the smack had knocked off into deeper water. The tug took her in tow in difficult conditions created by force seven winds from the NExE but it was clear that the casualty was sinking fast. Twice the lifeboat went alongside and took off a man each time but the third time the remaining crew would not leave even though by this time the **Lucky Hit** was awash. Soon after, the tow rope broke and she struck the North extension breakwater and sank. One man was washed overboard and was rescued by ropes thrown from the North Pier. Two others took to the rigging and were rescued by the lifeboat driving right over the submerged wreck and taking them aboard that way. The crew of five were: Charles Clarke – Skipper, H Mills – Mate, W Fish – Engineer, E Edmonds and G Lee – crew.
Later, in early March, the wreck was lifted clear of the harbour mouth by the salvage vessel **Cite de Londres** of Hull.

At the beginning of 1938 the London & North Eastern Railway Co took delivery of a new tug for Lowestoft - the **Ness Point**. She was a twin screw steam tug 71 feet long and of 85 tons gross. With the exception of Admiralty service during WWII she served at Lowestoft for 27 years. In 1965 she sank at her moorings and was raised and broken up later that year.

At the end of June 1938 Coxswain Spurgeon and his son William used their own small boat LT282 **Urge** to carry out a rescue. The *Agnes Cross* was away for her annual overhaul and with the Coastguards Albert Spurgeon had been keeping watch on two boys in a canoe. They had hoisted a sail and were soon in trouble with the strong ebb tide and heavy squalls. When it was seen that the boys were heading straight for the Newcome Sand the Coxswain and his son went off in their boat and brought the two back. The *Lowestoft Journal* of 2nd July gave their names as Graham Munday and Ernest Hutchinson.

In 1919, Cochrane of Selby had built the steam trawler **William Leech** which, at a later date, had been sold to French owners and renamed **Excellent**. In the early hours of Thursday 15th September 1938 she ran aground on the Newcome Sand in a strong NE breeze. The *Agnes Cross* was out at 4.40am but the Master refused help, saying he would wait for daylight. The lifeboat stood by and eventually the vessel refloated and continued her voyage.

A week later, on the 22nd, the *Agnes Cross* was out again this time looking for a crashed aircraft. All she found was a smoke float which, later, was thought to have fallen from an aircraft giving the impression that a plane was on fire and about to ditch.

Again, on Tuesday 27th September, the *Agnes Cross* was out to white smoke seen billowing, it was thought, from a vessel on fire to the SSE of the port. All that was found was a large barrel of coconut oil which was thought to have been disturbed by an explosion on the seabed (a World War I mine?).

At about 7.0am on Friday 25th November 1938 Jack Gennery, the Assistant Mechanic of the lifeboat, saw through the mist a steamer aground on the north-west part of the Newcome. Both the *Agnes Cross* and the tug **Lowestoft** went out. They found that the ss **Dunleith** of Poole bound for Edinburgh from Portsmouth had floated off. The lifeboat escorted her across the banks. Despite grounding she was found to be watertight and continued on her way.

During 1939, in the months leading up to World War II, there were only four calls for the lifeboat.

On Tuesday 3rd January four youths "borrowed" a dinghy and got into difficulties near the Inner Shoal. Albert Spurgeon was at the Coastguard lookout and saw the boat drifting with men in it signalling for help. The *Agnes Cross* was launched at about 11.35am in a strong NNW wind and heavy seas. When the lifeboat got alongside the boat the four of them jumped in and, together with their boat, were landed in the harbour. They all expressed their thanks for being rescued.

Later, the *Lowestoft Journal* of 18th February reported that three unemployed fishermen and a 15 year-old lad were charged with taking a boat from Waveney Dock and damaging it to the extent of £2. The court was told that the service had cost the RNLI about £20. The boat, a former ship's boat, was owned by Charles Henry Hutchinson of 1 Trinity Road, Beach village.

On the evening of Monday 29th May 1939 the pleasure boat **Joy**, with 14 passengers and three crew, suffered engine failure about a mile SSE of the harbour. Coxswain Spurgeon was told by the Coastguard at about 8.0pm and the ***Agnes Cross*** was away smartly at 8.10pm. They found the **Joy** with an overheated engine and took her in tow in the fresh north-east wind. They were safely in by 9.0pm.

On Monday 17th July 1939 the last service before the outbreak of war was carried out. At about 8.0pm the Coastguard received a message that a canoe had capsized off Ness Point in rough seas. The relief boat ***John & Mary Meiklam of Gladswood*** was away at 8.20pm and found a girl and a boy hanging onto the canoe and totally exhausted. Albert Spurgeon wrote in his log-book that they both had to be lifted into the lifeboat and treated for shock and cold. They were landed at the Yacht Club steps and both thanked the Coxswain for their prompt rescue.

Sixty years later, in 1999, the author had great pleasure in talking to both of them, still living in Lowestoft and got more complete details.
It seems that James Stanner, who was 19 years old, had got the canoe secondhand and had rigged it with a small shrimp trawl. He had gone out in calm conditions but off Ness Point, where it was choppy, the trawl came fast and he capsized. He could not swim and as fast as he clung to the canoe it rolled over.
Among the crowds watching from the beach was Ethel Steward aged 17. She told me that since no one else was doing anything she thought she had better. She remembered going into the water wearing a long skirt, which she dare not take off and that by the time she got to him she was exhausted. At least she was able to cling to the canoe opposite James and prevent it rolling over. She also remembered being roughly hauled into the lifeboat and the lifeboat crew thinking they were both mad to have gone out – not realising she was a rescuer.
Such was her courage for a 17-year old who had only recently gained her Girl Guide badge for lifesaving that she was awarded the Royal Humane Society Bronze medal. Ethel thought it was presented by Lady Somerleyton and she remembered being embarrassed by all the fuss. Later, during the war she was in the WRNS. She married Fred Wharton who later became a well known butcher in the town. Sadly he died in 2004.

James remembered The Mayor, Major SW Humphrey, being at the Yacht Club when they were landed and offering to pay for swimming lessons for him. Probably because of the outbreak of war the offer was not taken up. Jim served as an engine fitter in the Royal Navy during the war and then spent the rest of his working life at Eastern Coachworks in Lowestoft.

On Saturday 19th August 1939 the Dutch lifeboat ***Neeltje Jacuba*** from Ijmuiden paid a courtesy visit to the town. Crowds on the South front and the piers watched her arrival escorted by the ***Agnes Cross***. Various Dutch lifeboat officials were with her and they were welcomed by Cmdr. Vaux the Chief Inspector of the RNLI. Coxswain Albert Spurgeon gave the Dutch visitors a guided tour of the town.

Circumstances delayed the return visit by the Lowestoft lifeboat until August 1945.

1.3 - The Second World War and after

At the outbreak of hostilities, on 3rd September 1939, the Port of Lowestoft became a Naval Base and shipping only moved in or out of the harbour with the permission of the Senior Naval Officer i/c. The lifeboat could also only put to sea with permission. Reports on wartime rescues were very restricted or did not appear at all in the newspapers. However, with the help of a RNLI publication of 1946 kindly provided by Jeff Morris of The Lifeboat Enthusiasts Society, I have managed to get some details. In addition, some of the calls were to crashed aircraft and I have received much help and information from Bob Collis the well known local Aviation Historian.

The first wartime call for *Agnes Cross* was not long in coming. On Wednesday 13th September two vessels were in collision off the port. After much searching in a NE gale together with the Gorleston lifeboat, a destroyer, **HMS Wallace** was found, badly damaged but not in need of any help.

A month later on Thursday 12th October *Agnes Cross* carried out her last service at Lowestoft. Two longshore boats (LT1214) **Sonny Boy** and (LT259) **Joan** were flying distress signals. Help was in fact not needed, since by the time the lifeboat got to them one had been taken in tow by a drifter and the other expected to be able to restart her engine and so declined any help.

Whilst she had been at Lowestoft the *Agnes Cross* had launched 117 times and had saved 209 people.

The *Lowestoft Journal* of 28th October 1939 reported that the new lifeboat **Michael Stephens** had arrived at Lowestoft. Albert Spurgeon and three of his crew had travelled to Cowes, Isle of Wight to bring back the new boat. The passage was very rough and Coxswain Spurgeon told the paper that he was very pleased with the boat's performance. In fact, four new boats had left Cowes at the same time all destined for the east coast; the others going to Hartlepool, Tynemouth and Aberdeen.

The first service launch of ***Michael Stephens*** was on Sunday 12th November. A convoy of seven vessels had gone aground on the Newcome Sand in thick fog. The lifeboat launched at 4.30am and when the fog lifted six of them refloated on the rising tide but the ss **Appledore** remained hard and fast. The lifeboat guided the six to deeper water and then returned to the **Appledore**. Despite the efforts of two tugs she could not be shifted and the lifeboat returned to her mooring after eight hours out. Later in the evening the Senior Naval Officer required urgent letters to be taken to the Master of the **Appledore**. The lifeboat did this and was back by 11.30pm. The **Appledore** refloated the next day.

At the end of November Mr George Rose of Old Nelson Street died aged 72. He had been a crew member and latterly Second Coxswain for a total of 45 years. Six of his sons also served in the lifeboat. He had retired in 1936 but continued as a crew member on an "if needed" basis. His last service had been in the ***Agnes Cross*** in October and he had dearly wanted to sail in the new boat.

During the war it was often the case that explosions were heard out to sea. Sometimes it would have been mines going off, warships firing their guns or, worse, vessels being mined or attacked.

The first service of the war which resulted from enemy action came on Monday 12th December when the ss **Willowpool** of Hartlepool was set on fire 14 miles north east of Great Yarmouth. A total of 36 survivors had taken refuge on board the **Newarp** lightvessel; the Gorleston lifeboat ***Louise Stephens*** (ON820) took them off and landed them at Gorleston. The ***Michael Stephens*** launched at 1.15am and, in thick fog, made her journey by way of the **St. Nicholas**, **Cockle** and **Newarp** lightvessels to the burning **Willowpool**. The vessel was deserted and so the lifeboat lay alongside the **Newarp** lightvessel until daylight and then resumed searching. Nothing was found and she was back at Lowestoft by 10.0am. The Caister lifeboat ***Charles Burton*** (ON526) was also involved in this service.

On Saturday 23rd December the ***Michael Stephens*** was out for over 16 hours to help the motor trawler LT249 **Purple Heather** which was aground on Corton beach. At about 7.0pm the lifeboat was called on a very cold and foggy night. She found the trawler in amongst old sea defence works and iron pilings. Two crew were taken off, the others remaining aboard. The lifeboat stood by throughout the night. Coxswain Spurgeon thought it the bitterest weather in all his 35 years service with ice three inches thick on the deck and icicles hanging from the rigging. With daylight the lifeboat tried but could not refloat the trawler and since she was in no danger returned to harbour. The trawler was eventually refloated by a tug.

On Tuesday 30th January 1940 the ss **Royal Crown** of Hull was bombed by enemy aircraft and set on fire 15 miles south of **Smith's Knoll LV**. Some of her crew, in a ship's boat, came ashore at 7.0pm about eight miles south of Lowestoft at Easton Bavents. The Civil Defence report indicates that their boat capsized close inshore and seven of the 22 on board were drowned. The survivors were taken to Southwold hospital where they reported that another boat, with about 20 men, was still at sea. A whole ENE gale was blowing with heavy seas and squalls of rain and hail. At about 9.15pm the Naval Authorities passed a message to Coxswain Spurgeon that the boat was off Kessingland inside the Newcome and Barnard Sands. The *Michael Stephens* went off to this area being continually swept by icy seas. Finding nothing, Albert Spurgeon turned north and continued searching but without success. Since it seemed unlikely that a small boat would have survived such weather they returned to harbour at 1.45am on the 31st. The lifeboatmen got dry clothing and some food but remained on duty until at 5.45am the lifeboat put to sea again. She searched for several more hours but without finding any trace of the missing boat. The Southwold lifeboat *Mary Scott* (ON691) was also out for many hours and returned to her station at 7.0am. It was later assumed that the small boat must have capsized and all in her were lost.

At the end of May 1940 the historic evacuation of Allied forces from Dunkirk took place. Many "little ships" were to the fore in ferrying men off the beaches to larger vessels lying offshore. The *Michael Stephens* took part but with a Naval crew. The Lowestoft crew took the boat to Dover and collected her again when the operation was over. Branch records show that it cost £46.8s.7d to deliver and later collect the boat. However the RNLI did not charge the Government in respect of any of their boats used during the evacuation.

On Friday 4th October 1940 a German Junkers 88 bomber was shot down about 10 miles east of Southwold. Sqn.Ldr. R R Stanford-Tuck, flying a Hurricane of 257 Squadron based at Martlesham, claimed the hit though he thought the aircraft crashed close inshore. His log-book indicates that he circled the area for five minutes but seeing no sign of life he returned to base. The *Michael Stephens* was launched at 11.10am but found only oily patches on the water. The issue is unclear because at the same time a Heinkel 111 definitely did crash off the Suffolk coast and the bodies of the four crew members were washed ashore at Shingle Street. However, research by Bob Collis clearly points to the *Michael Stephens* being called to the crashed Ju88 where nothing was found.

On Sunday 29th December 1940 the Coastguard reported a British aircraft down in the sea about 3 miles ENE of Lowestoft. *Michael Stephens* was away at 12.30pm and searched for three hours but found nothing except a blue mitten which sank before it could be recovered. Bob Collis reports that no British aircraft fits this description though Sqn.Ldr Stanford-Tuck claimed a Dornier 17 to the south-east of Lowestoft. German records say non was lost that day. What, if any, aircraft it was will remain a mystery.

On Thursday 8th May 1941 the Naval Base reported a huge explosion where the Port Examination vessel had last been seen. She was HMT **Thistle**. The *Michael Stephens* launched at 9.10am into a fresh NE breeze and a sea strewn with mines. One of these blew up only 100 yards astern of the lifeboat. She found the wreck of the **Thistle** about two cables from the NE Newcome buoy and one very badly injured survivor out of a crew of 15. He was landed quickly and the lifeboat put out again amongst the mines but found no further survivors.

Another crashed aircraft claimed the attention of *Michael Stephens* on Tuesday 15th July 1941 but nothing was found. Bob Collis reports that a Wellington bomber from 214 Squadron took off from RAF Stadishall at 11.15pm (14th) but exploded and crashed soon after crossing the Suffolk coastline. The crew of the Wellington were:

>P/O VK Brown RAF Sgt. JS Else RAF
>Sgt. MR Collins RCAF Sgt. RD Hull RAF
>Sgt. J Taylor RAF Fl. Sgt. WG Lewis RAF

Pilot Officer Brown's parachute was washed ashore at Pakefield but non of the crew was found despite much searching by the lifeboat. Eventually the bodies of Lewis and Hull were washed ashore in England and that of Brown in German-occupied Holland.

During the night of Monday/Tuesday 8th/9th September the *Michael Stephens* had a long search for a missing fishing boat LT232 **Happy Days**. She had been reported at 7.30pm on the 8th flying a distress signal about 1½ miles SxW from Benacre and about 10 miles south of Lowestoft. Albert Spurgeon launched the lifeboat at 8.0pm and searched, in pitch blackness, until 11.30pm when he anchored. The search resumed at 4.0am (9th) and they found **Happy Days** and her crew of three at six o'clock with a broken down engine. They took the fishing boat in tow and were back at 8.0am.

Later in 1941 a recently formed sandbank at the harbour mouth started to cause problems. On Friday 21st November HMD **Rowantree** (ex BF199) stranded on that bank. The *Michael Stephens* launched at about 3.30pm and got a line aboard the vessel but it parted. It was left to a tug to tow her off but **Rowantree** capsized and the 15 crew managed to escape to the upturned hull. The lifeboat made two attempts at rescue getting 10 the first time and then the remaining five. The **Rowantree** had been built at Sandhaven in 1917. She had sailed at 8.15am and was returning to the port at about 3.0pm when she ran onto the bank.

On Christmas Day 1941 the *Michael Stephens* was called to rescue two men in a small boat drifting out to sea. She caught up with them about four miles south of the port. Their boat was half full of water and they were exhausted. Later, the *Lowestoft Journal* of 10th January 1942 reported that at Lowestoft Police Court, Ernest Albert Richard Wooden, unemployed, of 10 Kirkley Street was fined £10 with 10s.0d costs for navigating a rowing boat in tidal waters without permission. He claimed to have laid his lines the previous day without challenge. The youth with him was not prosecuted. The Bench took a serious view of this – no permission and hazarding the life of a young lad.

On Sunday 22nd February 1942 the ss **Enseign Marie Saint-Germain** of Newport, Mon was mined east of Caister, whilst in a convoy bound for the Tyne. The Coastguard asked for the Lowestoft lifeboat to help. The *Michael Stephens* was launched at 11.35am in a light northerly wind and five miles north of Caister, found two ship's boats with 25 survivors. They were landed at Yarmouth where it was learned that another boat with yet more survivors had returned to their ship which by now was under tow. Second Coxswain Arthur Swan was in charge of the lifeboat that day and he spoke to the steamer and finding that the remainder of the crew were safe on board he returned to Lowestoft at 6.25pm.

On Monday 26th July 1943 the *Michael Stephens* launched to reports of a crashed aircraft. Research by Bob Collis has indicated that a Short Stirling bomber from 75(NZ) Squadron took off from RAF Mepal in Cambridgeshire on a raid to Essen in Germany. At 2.38am the Royal Observer Corps unit at Aldeburgh reported that the Aldeburgh lifeboat had gone out to this aircraft which came down as it was approaching the Suffolk coast on its return. According to RNLI records it is unclear what the *Michael Stephens* did but it must be assumed that she searched also. The crew of the bomber were:

    Sgt MHC Ashdown RAF   Sgt RW Threadgold RAF
    Sgt RK Harrols RAF   WOII A Cleveland RCAF
    Sgt EC Denyer RAF   Sgt HC Dawson RCAF
    Sgt R Broadley RAF

The Aldeburgh lifeboat recovered the bodies of Dawson, Broadley, Ashdown and Denyer.

The only medal service of the war for the Lowestoft station took place on Thursday 30th September 1943.

On a dark and foggy night **HMMS 106** ran aground 1½ miles south of the harbour close to the Grand Hotel. At 9.40pm Albert Spurgeon had a message from the Naval Base to proceed to the minesweeper. Using the searchlight he was able to get alongside despite the vessel being on an old groyne and submerged iron pilings. The Commanding Officer asked for all the confidential books and equipment to be landed and this was done. A short time later two red rockets were seen and *Michael Stephens* put out again. To help, a searchlight was played on the harbour mouth but at a crucial moment it went out and in the sudden blackness the lifeboat hit one of the pierheads. Albert was thrown against the binnacle putting a three inch cut in his jaw but despite the Naval Officer on board saying he should return he refused. By the time they got back to the casualty she was listing and her crew and moveable equipment were taken on board the lifeboat and landed at the Naval Base at about 1.0am (1st Oct). Only then did Albert Spurgeon agree to go to hospital where his wound was stitched. Later that morning, at 11.0am, the lifeboat went off again this time under the command of the Second Coxswain Arthur Swan. However the wreck could not be towed off. The Officers and men of **HMMS 106** warmly thanked Albert Spurgeon and his crew for all their labours. Eventually it was decided to cut the bows of the minesweeper off and the rest of the vessel was towed into the harbour.

The minesweeper was eventually rebuilt and loaned to Italy. She was later renumbered 1606 in the Navy List and scrapped in 1956.

Much later the *Lowestoft Journal* of 31st May 1947 reported that Albert Spurgeon had been awarded the RNLI Bronze medal for this service.

After this very noteworthy service the remainder of the war was fairly quiet with only six calls in the remaining 20 months of war.

On Friday 11th May 1945 came two calls. The ss **Empire Dorritt** of Glasgow grounded on the Barnard Sand in foggy conditions. The *Michael Stephens* was launched at 5.25pm and stood by the vessel until she refloated on the tide and went on her way. The lifeboat was back at her mooring by 8.30pm only to be called again at 9.50pm. This time, the smack LT226 **Crecy** had run aground off Corton. She too refloated on the tide and the lifeboat towed her in at 11.30pm.

During the war the Lowestoft lifeboat answered 42 calls and saved the lives of 63 seamen.

Readers will remember that back in 1939 the Dutch lifeboat *Neeltje Jacuba* from Ijmuiden paid a courtesy visit to Lowestoft shortly before the outbreak of war. It was agreed, at the time, that the Lowestoft boat would return the visit though I expect that most people realised it would be some years away. The *Lowestoft Journal* for 25th August and 1st September 1945 carried reports about the visit.
Coxswain Spurgeon, 3 crew and Cmdr.Vaux (Chief Inspector of Lifeboats) set off in the **Michael Stephens** on Sunday 26th August towing the Dutch lifeboat **Jorn Hodoshen**. This vessel and a group of refugees had escaped from Holland in early 1944 under cover of darkness. They broke down half way across the North Sea and were found by a Patrol Service vessel and brought to Great Yarmouth where the boat remained until this voyage back to Holland.
They journeyed via Harwich and then to the Hook of Holland on a voyage of 21 hours against wind and sea. Albert and his men found a meal at the Marine barracks and a bed on a former German hospital ship. Next day they hitched a lift in an army lorry to Amsterdam where he reported that the shops had nothing to sell and trade was by barter. They moved on to Schiedam, had a look round and tried to find a lift to the Hook. Albert went

on to say, "Finally we found an English billet with a good meal and lashings of tea. Eventually we got a lift to Rotterdam and onwards to the Hook."
They left the following day into a WNW gale which meant they had a head wind and sea just as they had had on the outward trip.

The first call of 1946 came on Sunday 20th January when the **MFV 1165** ran onto Corton beach in a strong NNE wind. The *Michael Stephens* was away at 10.50pm but could not get alongside owing to the shallow water. Three of the casualty's crew were rescued by Coastguard rocket apparatus, three others having got ashore unaided. The RNLI reported that the vessel became totally wrecked but the *Lowestoft Journal* of 26th January reported that she took little harm and was refloated by the Yarmouth tug **Richard Lee Barber**. This seems most likely since **MFV 1165** was in the Navy List until 1954.

The *Eastern Evening News* of 29th August 1946 had the headline – "Lifeboat out with woman in the crew." On Wednesday the 28th the 10 ton yacht **Ziska** of Whitby was homeward bound from Ostend with a crew of seven. Off Lowestoft in a SSE gale and heavy seas she was in danger of being driven onto the Newcome Sand. Since time was pressing, Albert Spurgeon called for volunteers rather than wait for some of his regular crew. Two came forward - a Mrs Gooch of Leicester who was staying at the Victoria Hotel and a Mr Bagshaw, a local man. A tug had earlier tried to find the yacht off Southwold but without success. The *Michael Stephens* launched at 10.55am and eventually chanced to see the yacht's mast as it was being tossed about near the Newcome. She was waterlogged and the crew were baling. A tow was fixed up and the vessel was safely towed to harbour. The *Lowestoft Journal* of 31st August reported that Mrs Gooch was experienced in seamanship and thoroughly enjoyed the voyage.

On Friday 20th September 1946 the Lloyds Agent at Lowestoft asked if the lifeboat could take food out to a vessel anchored near the Inner Shoal buoy. The weather was too bad for a tug to go. The relief boat *Mary Scott* left at 5.25pm in heavy seas and a strong SSW wind. She found the ss **Brightside** of Middlesborough with a cargo of timber from Bergen for Lowestoft. The Master explained that the engine was disabled and that they had had no food for two days. Coxswain Spurgeon warned him that they were in a dangerous position and said that the shore would keep the vessel under observation and that the lifeboat would return if needed. At 3.0am on the 21st it was seen that the steamer was dragging her anchor and the lifeboat crew were assembled. In the event they were not needed. The *Lowestoft Journal* of 28th September reported that tugs stood by overnight and that towards noon on the 21st the **Brightside** made her own way into port with a list caused by the deck cargo having shifted.

At 4.30pm on Tuesday 17th December 1946 the Coastguard reported SOS messages from the trawler LT1239 **Helping Hand** which was under tow by the Grimsby trawler GY1281 **Brisbane**. The casualty was taking water via a broken propellor shaft and immediate help was needed. The *Michael Stephens* left at 5.45pm and found both vessels a mile from the East Newcome buoy. The lifeboat stood by both of them throughout a most uncomfortable night with the lifeboat being swept by heavy seas, until the tug **Lowestoft** arrived at 5.0am (18th) and towed **Helping Hand** into Lowestoft harbour where she was beached. William Ball, the Skipper of **Helping Hand**, thanked the lifeboat for standing by in what Coxswain Spurgeon described as the worst weather he could remember.

The following night the *Michael Stephens* was out again, this time to the London tanker **Esso Ottawa**. She had suffered engine failure and had anchored off the port with 85000 gallons of paraffin from Sheerness for Lowestoft. The lifeboat, under Second Coxswain Arthur Swan (Albert Spurgeon was laid up after the previous night's call), took off the crew of nine since the tanker was by then dragging her anchor in an easterly gale. After the tanker crew had had a meal they went out in the tug **Lowestoft** but failed to find their vessel in the darkness. They were put back aboard at first light on the 19th and the tug brought the tanker into the harbour.

The winter of 1946/47 was extremely severe and on one service Albert Spurgeon reported spray freezing all over the boat. This was on Monday 10th February 1947 when the ss **Cambrian Coast** of Liverpool, from Newcastle for London with coal broke down off Lowestoft. By the time *Michael Stephens* got to her she was seven miles outside the **Corton** lightvessel and needed a tow. The casualty's radio was out of action and the lifeboat could not contact the tug and so she returned to Lowestoft with various messages for the ship's owners and the tug. Eventually the Yarmouth tug **Richard Lee Barber** found the steamer and towed her into Yarmouth Roads

The *Lowestoft Journal* for 15th February 1947 reported worries from trawler owners that the northern approaches to Lowestoft were silting up because the

Newcome Sand was gradually creeping northwards. A new channel was opening to the south but Trinity House did not agree that this needed navigation buoys. Vessel owners were concerned that their ships would run aground either on the extended Newcome Sand or in the un-buoyed southern channel.

On Tuesday 3rd July 1947 the ex-German yawl **Helgoland**, attached to **HMS Ganges** the shore base at Shotley near Ipswich, grounded a mile SSE of the Coastguard station. The *Michael Stephens* was out by 4.25pm and stood by until she refloated and then towed her into the harbour. The records show that, by 1951 this yawl was owned by the Royal Naval Sailing Association and was named **Pickle** after **HMS Pickle**, the RN schooner which brought home the first news of the Battle of Trafalgar in October 1805.

Albert Spurgeon's RNLI service at Lowestoft ended in 1947 with a service to a crashed aircraft. At about 7.0am on Wednesday 24th September 1947 an RAF Avro Lincoln bomber of 97 Squadron based at Hemswell, Lincolnshire crashed ashore at Mautby near Yarmouth. The lifeboat was asked to search at sea for any crew who may have baled out. The *Michael Stephens* left harbour at about 7.30am and searched over a large area but found nothing. The Caister and Gorleston lifeboats also searched as did two aircraft.
The *Lowestoft Journal* of 27th September reported that, rather poignantly, one of the lost airmen Gunner II Roy Trundle, was a Lowestoft man the son of a baker, whose Mother then lived in London Road South. He was one of the early members of the Lowestoft Air Training Corps in which he became a Flight Sergeant. He was called up in 1943 and served in Bomber Command. He volunteered for another three years after the war, and was 22 years old when he died. Of the nine crew of this aircraft he was the only one identified. He is buried in Lowestoft cemetery and his eight comrades were interred in a group grave at Caister-on-Sea. There is a memorial to this crash in Mautby church.

So Albert Spurgeon's long contribution to the RNLI at Lowestoft ended. During his years in charge of the boats they had been launched 163 times and had saved 183 lives.

Also at the end of 1947 the Second Coxswain Arthur Swan retired after 49 years lifeboat service.

Table 1

Services under Coxswain Spurgeon 1924-1947

1925   *Agnes Cross*

29 Jan	ss **Taunton** of Grimsby, Corton	stood by
8 Feb	smack, **Ebenezer** (LT784), Newcome	saved 5
16 Apr	airship, **R33**, North Sea	recalled
1 May	dredger, **Mudsucker**, harbour mouth	stood by
19 May	longshore boats in squall	stood by
27 Jun	smack, **Irene** (LT976), N beach	saved 4
18 Oct	rowing boat, **Myra** adrift	saved 2 + boat
20 Oct	drifter, **Diligence** (FR917), N beach	stood by
23 Dec	ketch, **Henrietta** of Goole, Corton	ns

1926   *Agnes Cross*

6 Jan	smack, **Ivan** (LT1098), Newcome	stood by
3 Apr	gig from **HMS Godetia**, N beach	stood by
14 Apr	trawler, **Antonia Isabella** (O.65) of Ostend	assembled
25 Aug	trawler, **Arthur** (O.171) of Ostend, Corton	stood by
4 Oct	trawler, **Pattrick** (O.164) of Ostend, Newcome	stood by
13 Nov	ss **Agnes** of Haugesund, Newcome	stood by

1927   *Agnes Cross*

21 Feb	**Emblem** (no details)	assembled
23 Jul	trawler, **Yolande** (O.30) of Ostend, N beach	took off 4
2 Oct	flares	ns
10 Oct	very high tide and gales	assembled
24 Oct	**Johanna Marie** of Scheveningen, Newcome	assisted
21 Nov	smack, **Lily of Devon** (LT96), S beach	saved 3
29 Nov	smack, **Dusky Queen** (LT895), N extension	assisted
17 Dec	smack, **Sweet Home** (LT1269), Newcome	assembled
19 Dec	ss **Gunhild** of Copenhagen	ns
27 Dec	smack, **Wide Awake** (LT37), dismasted	stood by

1928 *Agnes Cross*

13 Jan	trawler, **Narva** of Ostend, N roads	stood by
15 Jan	smack, **Colinda** (LT382), N beach	saved 4
13 Apr	ss **Northam**, Inner Shoal	assembled
22 Jun	dinghy, **Ugly**, Inner Shoal	saved 1 + boat
23 Jun	schooner, **Mary Ann** of Jersey	assembled
24 Jun	trawler, **Godsgenard** (O.57) of Ostend, Newcome	assembled
25 Jun	fb, **Mica** (LT683), on fire	assisted
26 Aug	ss **E Rose** of Great Yarmouth, N beach	saved 8
1 Sep	smack, **Trio** (LT1272), on fire	ns
6 Nov	drifter, **Generous** (LT1148), N beach	ns
7 Dec	derelict, **Kate**, near Corton LV	ns
30 Dec	trawler, **Elnet** (IJM.80) of Ijmuiden, N beach	stood by

1929 *Agnes Cross*

30 Jan	smack, **San Toy** (LT1220), Pakefield	stood by
3 Feb	trawler, **Newhaven** (LT134), Newcome	stood by
18 Apr	dinghy, **Boy Fred**, Newcome	saved 3 + boat
21 Apr	smack, **Try On** (LT176), N extension	stood by
4 Oct	smack, **Kestrel** (LT1097), N beach	assisted
4 Oct	smack, **Shrublands** (LT1223), N beach	stood by
8 Nov	trawler, **Strive** (LT766), N beach	stood by
8 Nov	ss **Mallin** of Haugesund, N roads	ns
18 Nov	ss **Mile End** of London, Holm	stood by
5 Dec	drifter, **Reward** (LT463), Holm	ns
13 Dec	trawler, **Dinorah** (GY1107), burning flares	ns

1930 *Agnes Cross*

23 Mar	ss **Greenawn** of Goole, Outer ridge	stood by
7 May	lugger, **Enterprise**, Barnard	saved 2 + boat
25 May	rowing boat, **Tom Bowling**, Newcome	saved 3 + boat
5 Oct	fb, **Tennessee** (LT146), Inner Shoal	saved 2 + boat
8 Oct	**Gellyburn**, (no details)	assembled
24 Nov	drifter, **Impregnable** (LT1118), N extension	assembled
24 Nov	smack, **Kestrel** (LT1097), lost sails	assembled
24 Nov	drifter, **Adriana I** (VL.88). Inner Shoal	stood by

1931 *Agnes Cross*

26 Oct	drifter, **Bob Read** (LT2), N roads	assembled
18 Dec	lugger, **Boy Reggie** (LT104), Holm	saved 2 + boat

1932 *Agnes Cross*

13 Jan	barge, **Servic** of London, Corton	stood by
7 Apr	fb, **John & George** (LT155), Newcome	saved 1 + boat
12 Jun	barge, **Scone** of Rochester, on fire	assisted
13 Jun	lugger, **Boy Reggie** (LT104), ashore Southwold	assembled
28 Oct	lugger, **Sonny Boy** (LT1214), Newcome	saved 3 + boat
1 Nov	drifter, **Bezaleel** (K27), N beach	stood by

1933 *Agnes Cross*

16 Feb	smack, **Ivanhoe** (LT1271), S beach	stood by
31 Mar	m.cruiser, **Francomme**, Inner Shoal	saved 3 + boat

City of Bradford I

8 Jun	ss **Effra** of London, Ness Point	ns

Agnes Cross

21 Jun	two canoes, Kessingland	saved 2 + dog
14 Sep	trawler, **Flag Jack** (LT1224), Outer Bank	stood by
7 Dec	smack, **WEH** (LT910), N extension	assisted
7 Dec	fb, **Marjorie** (LT379), Newcome	saved 2 + boat

1934 *Agnes Cross*

1 Jul	ss **Marion Traber**, Corton	stood by
28 Aug	dinghies capsized, harbour mouth	assembled
18 Dec	barge, **Davenport** of Ipswich	stood by

1935 *Agnes Cross*

30 Jan	ss **Pagasitikos** of Andros, Newcome	stood by
17 Feb	small boat, Holm	ns

City of Bradford I

14 May	fb, **Joan** (LT259), Newcome	saved 2 + boat

1935 (cont)

Agnes Cross

30 Aug	dinghies, **Whimbrel** and **Snipe** capsized	saved 4 + 2 boats
24 Sep	yacht, **Magpie** of Burnham on Crouch, Newcome	saved 3 + boat
20 Oct	smack, **Challenger** (LT975), N extension	saved 4 + boat
3 Nov	fb, **Golden Miller** (LT83), Newcome	saved 2 + boat
13 Nov	fb, **Sonny Boy** (LT1214), Newcome	escorted in
13 Nov	drifter, **Gowan Bank** (BF193), Inner Shoal	stood by
30 Dec	trawler, **Rochester** (LT153), Inner Shoal	ns

1936 *Agnes Cross*

| 23 Feb | *RNLB Baltic*, overdue | towed in |

J B Proudfoot

| 17 May | barge, **Will Everard** of London, Scroby | stood by |
| 25 May | gunfire – warships exercising | ns |

Agnes Cross

28 Jul	fire at sea	ns
3 Aug	rowing boat, Newcome	saved 4 + boat
11 Sep	trawler, **Blarney Rose** (LT1242), S beach	ns
4 Oct	hydroplane, **Nine**, Ness Point	ns
5 Oct	barge, **Cetus** of London, Newcome	stood by
20 Oct	smack, **Concord** (LT263), Barnard	ns
15 Nov	m cruiser, **Orinthia** on fire, N beach	ns
12 Dec	flares, Barnard	ns

1937 *City of Bradford I*

| 10 May | m boat from **HMS Fitzroy**, Ness Point | saved 8 + boat |
| 17 May | m boat from **HMS Fitzroy** | stood by |

Agnes Cross

1 Nov	fb, **Terry** (LT379), Pakefield	saved 3 + boat
7 Nov	ss **Ascania** of Wiborg, Newcome	ns
9 Nov	fb, **Joan** (LT259), Barnard	saved 2 + boat
1 Dec	drifter, **Pilot Star** (KY48), Holm	escorted in
6 Dec	smack, **Try On** (LT176), Inner Shoal	escorted in

1938   *Agnes Cross*

8 Feb	drifter, **Lord Keith** (LT181), Benacre	assisted
14 Feb	smack, **Northern Queen** (LT1099), lost sails	stood by
16 Feb	ss **St. Helena** of Finland, North Sea	ns
18 Feb	smack, **Lucky Hit** (LT961), Newcome	saved 4
8 Apr	yacht, **Eleanor** of Boston, Newcome	saved 2 + boat
30 May	yacht, **Merry Thought** of Newhaven	escorted in
15 Sep	trawler, **Excellent** (B.915), Newcome	stood by
16 Sep	yacht, **Seagull**, Kessingland	saved boat
22 Sep	aircraft in sea	ns
27 Sep	fire at sea	ns
23 Nov	2 barges in gale	ns
25 Nov	ss **Dunleith** of Poole, Newcome	stood by
20 Dec	unknown Dutch vessel, no details	ns

1939   *Agnes Cross*

3 Jan	yacht, **Roustabout**, Inner Shoal	saved 4 + boat
17 Jan	unknown trawler	ns
29 May	m launch, **Joy**, engine failure	saved 17 + boat

John & Mary Meiklam of Gladswood

17 Jul	canoe, Ness Point	saved 2 + canoe

Agnes Cross

13 Sep	**HMS Wallace**	ns
12 Oct	fb's, **Sonny Boy** (LT1214) & **Joan** (LT259)	ns

Michael Stephens

12 Nov	ss **Appledore**, Newcome	escorted
12 Nov	ss **Appledore**, delivered letters from SNO	
25 Nov	explosions	ns
4 Dec	explosions	ns
8 Dec	**HMS Susette**, N beach	assembled
9 Dec	**HMS Susette**, N beach	stood by
12 Dec	ss **Willowpool** of Hartlepool, on fire	ns
20 Dec	aircraft in sea	ns
21 Dec	**HMS Saxonia**, Newcome	stood by
23 Dec	trawler, **Purple Heather** (LT249), Corton	stood by

1940   *Michael Stephens*

5 Jan	trawler, **Adrian** (LT114)	stood by
16 Jan	barge, **Britisher** of London, lost sails	stood by
30 Jan	ship's boats from ss **Royal Crown** of Hull	ns
9 Apr	boat from **HMS Waverley**, Newcome	saved 3 + boat

May/June   at Dunkirk

11 Aug	aircraft in sea	ns
4 Oct	aircraft in sea	ns
21 Dec	**HMT Niblick**, Newcome	escorted in
29 Dec	aircraft in sea	ns

1941   *Michael Stephens*

8 May	**HMT Thistle**, sunk by mine	saved 1
15 Jul	aircraft in sea	ns
6 Aug	6 steamers on Haisboro Sand	ns
8 Sep	fb, **Happy Days** (LT232), Benacre	saved 3 + boat

J B Proudfoot

3 Oct	fb's, **Winston** (LT142) and **Ranger** (LT855)	ns

Michael Stephens

21 Nov	fb, **Joy** (LT337)	saved 4 + boat
21 Nov	**HMD Rowan Tree**, capsized	saved 15
25 Dec	rowing boat, S roads	saved 2

1942   *Michael Stephens*

9 Feb	**HMD Golden News**, Newcome	stood by
22 Feb	ss **Enseign Marie Saint-Germain** of Newport	saved 25
19 Apr	fb, **Boy Eric** (LT1241)	saved 2 + boat

1943   *Michael Stephens*

4 Mar	motor gun boat	ns
26 Jul	aircraft in sea	ns
30 Sep	**HMMS 106**, Pakefield beach	saved 10

1944	*Michael Stephens*	
10 Feb	small boat, Kessingland	ns
1945	*Michael Stephens*	
5 Feb	smack, **Northern Queen** (LT1099)	assembled
31 Mar	aircraft in sea	ns
29 Apr	fb, **Unity** (LT428), Southwold	escorted in
11 May	ss **Empire Dorritt** of Glasgow, Barnard	stood by
11 May	smack, **Crecy** (LT226), Corton	assisted
31 May	**HMT Alexandrite**, Holm	stood by
15 Jun	aircraft in sea	ns
9 Sep	fb, **Viking** (LT11)	ns
	Mary Scott	
2 Oct	drifter, **Covent Garden** (LT1258), Inner Shoal	stood by
1946	*Michael Stephens*	
20 Jan	**MFV1165,** Corton beach	stood by
13 May	mv **Loa Ronn** of Denmark, Newcome	assisted
9 Jun	canoe, Ness Point	ns
28 Aug	yacht, **Ziska** of Whitby, Newcome	escorted in
6 Sep	small boat, Inner Shoal	ns
	Mary Scott	
20 Sep	ss **Brightside** of Middlesborough, Inner Shoal	took out food
	Michael Stephens	
27 Oct	sailing boat, **Chums**, Inner Shoal	saved 4 + boat
12 Nov	fb, **Don't Know** (YH422), Inner Shoal	saved 3 + boat
17 Dec	trawler, **Helping Hand** (LT1239)	stood by
18 Dec	mv **Esso Ottawa** of London	took off 8

1947   *Michael Stephens*

15 Jan	ex HMML, Covehithe	stood by
10 Feb	ss **Cambrian Coast** of Liverpool	stood by
11 Mar	trawler, **Isle of Wight** (H852), Newcome	assembled
3 Jul	yawl, **Helgoland** of Shotley, Holm	assisted
9 Jul	ss **Sheaf Crown** of Newcastle	stood by
16 Sep	yacht, **Elsquir**, Inner Shoal	escorted in
24 Sep	ditched airmen, search for	ns

Total 163 launches, 183 people plus 1 dog saved.

Coxswain Albert Spurgeon, he was in charge of Lowestoft lifeboats from 1924 until 1947.
(Lowestoft Maritime Museum collection)

RNLB *Agnes Cross* (ON663), stationed at Lowestoft from 1921 until 1939. (PLRS collection)

The sailing smack **Irene** (LT976) aground on the North beach in June 1925. (PLRS collection)

The Dutch steam trawler **Elnet** (IJM.80) aground on the North beach near the Coastguard lookout in December 1928. (PLRS collection)

The lifeboat houses on the North beach, built in 1858 (right) and 1870 (left). They were demolished in 1972. (PLRS collection)

LOWESTOFT

HOBSON & CO. (LOWESTOFT) LTD.

Have received instructions from the Royal National Lifeboat Institution to sell by Auction on

Wednesday May 20th, 1931

THE LATE WINTERTON

LIFEBOAT

(NORFOLK AND SUFFOLK TYPE)

LENGTH 44 ft. 6 ins. BREADTH 12 ft. 6 ins.

Clinker Built. Mahogany Planking Copper fastened. Timbers of Canadian Rock Elm. Stem Post and Deadwoods of English Oak. Weight of Iron Keel $22\frac{1}{2}$ cwts. Two Drop Keels weigh about $2\frac{3}{4}$ cwt. each. Together with Masts, Sails, Rigging. Anchor and Cable. Date of construction 1909.

The Auction to take place at

LIFEBOAT SHED, BEACH. LOWESTOFT,

(Where the above Lifeboat is stored), at

TWELVE O'CLOCK NOON PRECISELY.

For further Particulars and for Conditions of Sale apply to the

AUCTIONEERS, The Gables, Marina, Lowestoft.

Flood & Son, Ltd., The Borough Press, Lowestoft.

A poster advertising the sale of the former Winterton lifeboat RNLB *Eleanor Brown* (ON589) by Hobson & Co in 1931. (PLRS collection)

The trawler **Lucky Hit** (LT961) sank in the harbour mouth in February 1938. The *Agnes Cross* is seen here preparing to rescue two men who are in the rigging. (PLRS collection)

The motor trawler **Purple Heather** (LT249) went ashore at Corton in December 1939. The *Michael Stephens* stood by in bitter weather all night, the trawler being refloated later in the day. (PLRS collection)

The ss **Brightside** of Middlesborough broke down off Lowestoft in September 1946. She came into port listing because of a shifted deck cargo of timber. (PLRS collection)

The London tanker **Esso Ottawa** suffered engine failure off Lowestoft in December 1946. She was eventually towed into port by the tug **Lowestoft**. (PLRS collection)

CHAPTER 2

Coxswain Burgess 1947 – 1968

2.1 – Mid 20th Century

On Wednesday 1st October 1947 Harry Thomas Burgess was appointed Coxswain of the Lowestoft lifeboat; at that time, aged 37, he was the youngest on the east coast. He was the Grandson of George Rose who was Second Coxswain for many years. In succession to Arthur Rose the new Second Coxswain was Mr C Rose who was an uncle to Harry Burgess.

Within a week, on the 7th, he was out helping the Grimsby trawler GY441 **Ireland's Eye** which grounded near the North pier. The lifeboat ran out an anchor from the vessel and then stood by. Eventually the tug **Lowestoft** got the trawler off.

On Thursday 29th April 1948 the Coastguard reported a dinghy drifting off Corton with a person on board. The *Michael Stephens* got out to it but found that the 'person' was in fact two bushes stuck in the boat.

On Thursday 21st July 1948 the motor cruiser **Dimcyl**, on a voyage from Lowestoft to Holland, broke down off Orfordness. She returned to Lowestoft under sail but her owner and her crew of four decided not to attempt to enter the port without an engine. Harry Burgess launched the relief boat *Mary Scott* into a strong southerly wind and towed the boat in by 4.30pm.

At about 8.45pm on Monday 13th September 1948 the Coastguard reported that a vessel was burning flares about six miles SSE of the port. Harry Burgess took the *Michael Stephens* in a fresh south westerly wind and found the Belgian trawler N.814 **Roger** broken down. She had been fishing near the Outer Gabbard but on the Saturday the weather became so bad that she was forced to run for shelter. By the Monday her engine was giving trouble so she anchored and eventually burned distress signals. The *Michael Stephens* towed her into Lowestoft at nearly midnight.

On New Year's Day 1949 the Lowestoft trawler LT245 **JAP** was on her way home from the fishing grounds. She touched the bottom near the Inner Shoal and damaged the rudder making her difficult to control. Her signals of distress were seen and the *Michael Stephens* launched at about 9.15pm. Harry Burgess found that she was not in any immediate danger and returned for a tug. The lifeboat stood by until the tug **Lowestoft** got her off and into the harbour at about 2.0am (2nd).

On Wednesday 4th May 1949 the Dutch motor coaster **Mudo** sank after colliding with the tanker **Algol** about four miles ESE of Lowestoft. It was very foggy and the sea was calm when the *Michael Stephens* launched at 7.45pm. She searched until 11.0pm but found neither of the two men lost from the **Mudo**. The *Lowestoft Journal* of 6th May reported Harry Burgess as saying, "We had a hopeless task in the thick fog and darkness. Although we kept a good lookout we saw no sign of the men. All we saw was oil on the water."

On Saturday 14th May 1949 the ss **PLM 17** of Rouen grounded on the Newcome. Harry Burgess took the *Michael Stephens* out at 2.15am and put one of his crew men aboard the casualty. The lifeboat then stood by until the steamer was refloated by the Yarmouth tug **Richard Lee Barber**. The **PLM 17** was of 3754 tons, owned by French National Railways and was bound to the Tyne from Sfax in Tunisia. The lifeboat was finally back at her station at 1.0pm after 11 hours out.

On Sunday 27th November 1949 the Yarmouth drifter YH344 **Ocean Sunbeam** ran aground on the Newcome in a northerly wind and heavy swell. The *Michael Stephens* stood by for an hour or so until the drifter got off under her own power at about midday.

The new-year 1950 was quiet for the lifeboat with only five calls all year.

On Tuesday 9th May the trawler LT309 **Loddon** stranded on the Inner Shoal in thick fog. The *Lowestoft Journal* of 12th May reported that the vessel, owned by Consolidated Fisheries, sounded distress signals on her siren. Harry Burgess took the *Michael Stephens* out at 4.0pm and stood by the casualty for an hour and a half until the tug **Ness Point** got her off and she was able to get in under her own power.

On Monday 11th September 1950 the motor yacht **Heron** of Rochester was reported in difficulty off Southwold. The *Michael Stephens* went out. The owner, Mr CW Bacon of the Little Ship Club, reported engine failure but was able to make a little headway. The lifeboat escorted her to Lowestoft.

On Sunday 17th December 1950 the ss **Rocquaine** of Guernsey was four miles off Lowestoft and unable to raise steam because the boiler crown had collapsed. The *Michael Stephens* launched at 5.30am and stood by in a fresh WNW wind and snow showers until the Yarmouth tug **Richard Lee Barber** was able to tow the casualty to Yarmouth. Harry Burgess's log book recorded that the vessel was en route from Goole to Guernsey with general cargo.

On Saturday 10th March 1951 the *Michael Stephens* stood by while the tug **Lowestoft** towed in the trawler LT160 **King Athelstan**. Three days earlier she had broken down with boiler problems and her sister ship LT309 **Loddon** had towed her the 90 miles to Lowestoft. Off the harbour she broke adrift and went on the Newcome. She was owned by Consolidated Fisheries and was one of the older vessels in the port having been built in Grimsby in 1899. The *Lowestoft Journal* of 12th March reported that her Skipper, John Attridge had, the year before, been awarded the Royal Humane Society's Bronze medal for saving the life of one of his crew who had fallen overboard.

On Saturday 5th May 1951 Second Coxswain Charles Rose was in charge when the *Michael Stephens* went to a small Yarmouth shrimper which sank off Corton. The two crew clung to the mast and a 14-year old schoolboy, Michael Walker of London who was staying at Corton Holiday Camp, swam out with a lifebelt by which means the two were hauled ashore. The shrimp boat was owned by Charles Simmons of Gorleston and his son Charles. It was thought that their boat sank after striking old wartime anti-invasion defences. They were safely ashore by the time the lifeboat arrived.

On Sunday 11th November 1951 the longshore boat LT381 **Wavell** was heading back to Lowestoft with a huge catch of sprats when the engine failed. By the time Jack Hales and Charlie Button got it started again they had drifted as far as Southwold. They set off north again but off Pakefield the engine stopped. In the darkness and with such a heavy load of sprats they were taking water which kept them pumping for three hours even though they were anchored. Eventually they sent up flares and both the

Michael Stephens and the Coastguard Lifesaving Apparatus team were called. The lifeboat got a line to the **Wavell** and towed them in by 9.30pm. The *Lowestoft Journal* of 16th November reported the catch as 150 maunds – the biggest single catch since the war.

On Thursday 17th January 1952 the Dutch mv **Seaham** was twelve miles ESE of Lowestoft in a NW gale when her cargo of coal, destined for Rotterdam, shifted. Four of her crew took to a small boat which was capsized and three were lost, the fourth being picked up by the Trinity House Vessel **Warden**. The *Michael Stephens* was launched at 4.15pm and when she reached the casualty she was able to take off three engineers leaving the Master to try to connect a tow. This was unsuccessful and Harry Burgess went alongside a second time to take off the Master and later collected the rescued man from the **Warden**. The lifeboat suffered some damage at this time to the wireless mast and handrails. The *Michael Stephens* was back at 12.15am (18th).

An interesting variation on ways of calling out the lifeboat happened on Thursday 1st May 1952. Police Constable Simmons was sitting on the top deck of a Lowestoft Corporation bus which was crossing the swing bridge. At that moment the Constable saw a sailing dinghy capsize off the harbour mouth. He left the bus and telephoned the Police Station who called the Coastguard. The maroons were fired and the relief lifeboat *Mary Scott* was away at 7.15pm. She caught up with the dinghy, being carried north on the tide, off the lighthouse. Two men, Mr Rodney Grantham-Hill of Beccles and Dr Anthony Wells of Oxford were hanging on to her, one with his feet trapped in the rigging. The *Lowestoft Journal* of 2nd May reported that Harry Burgess said, "We eventually managed to free him and get him into the boat where artificial respiration was given. He was lucky to be alive." On landing they were both taken to Lowestoft Hospital.

A broken down fishing boat SH96 **Monbretia** led to the *Michael Stephens* being called on Thursday 20th November 1952. The vessel was on her way from Yarmouth to Brightlingsea when her engine failed. She was initially taken in tow by the mv **Novian Coast**, on passage to Middlesborough but the tow parted and the fishing boat drifted towards the Newcome Sand in a fresh southerly wind. The lifeboat under Harry Burgess launched at 1.40pm and found the **Monbretia** three miles off Benacre Ness. She was taken in tow and handed over to the tug **Ness Point** off the harbour. A salvage award of £90 was made to the lifeboat. The fishing boat was eventually able to leave for Brightlingsea on 2nd December.

On Tuesday 9th December 1952 the motor trawler LT123 **Mollia** grounded several times off the port in thick fog. Harry Burgess heard a vessel sounding her siren for help and went to the Coastguard lookout. Eventually the fog thinned and he saw flares about 1½ miles to the SSE. The *Michael Stephens* was launched about 5.0pm and went to the casualty. Skipper S Coble of the **Mollia** thought he must have mistaken the light on the Claremont Pier for the harbour entrance after having run aground earlier. He had decided to anchor after the latest mishap and fired flares for help. His ship was towed in by about 7.0pm.

The first few months of 1953 were very quiet for the lifeboat but when the whole country was celebrating the Coronation of Queen Elizabeth II there came a call!! The *Eastern Evening News* of 2nd June 1953 described it as "A Coronation Day dash."
The fishing boat LT50 **Marion,** with three crew on board, fouled her propeller while fishing off Pakefield. She was seen by the Coastguard flying a distress signal about three miles SSW of the harbour in a fresh NNW wind. The *Michael Stephens* was away at 12.25pm and found the boat with a trawl, which had fallen over the side, wrapped round the propeller. The owner, Mr J Gaze, said he had no sails and therefore he anchored. A lifeboat crew member was put aboard to help and the lifeboat towed the **Marion** home arriving at 2.40pm. Mr Gaze later sent £3 for the lifeboat.

On Saturday 1st August 1953 flares were seen about seven miles south of Lowestoft. The *Michael Stephens* put out at 11.50pm and found the tug **Armina** burnt out and abandoned. She was burnt to the waterline and no person could have survived aboard her. The lifeboat searched the area and found a ladder and oil drums rigged as a raft. They later found three men in a small boat but it transpired that they had set off from Kessingland to help, had not realised how far out the casualty was and were exhausted by the effort. They were towed to the beach and the lifeboat returned to the harbour at 3.50am (2nd). Eventually a wireless message was received saying that the sole occupant of the tug had been picked up by mv **Hudson Firth**. The owner of **Armina**, Mr Alan Delf of Corton Road, Lowestoft, explained that he was bringing her back from Wivenhoe when she caught fire off Southwold. He had used extinguishers and water to no effect. The *Lowestoft Journal* of 7th August recorded that he had rigged his raft and abandoned ship with two lifejackets. He had drifted some distrance before **Hudson Firth** found him. The **Armina** was a small ex-US tug which Mr Delf used for towing yachts on delivery runs.

On Sunday 22nd November 1953 the small fishing boat LT320 **Belle** with three on board developed engine trouble off the harbour. Harry Burgess who was watching from the Coastguard lookout decided to launch the *Michael Stephens* since the boat was rapidly driving north. He caught up with them two miles north of the harbour and found they had no sails and the anchor cable was too short for it to hold on the bottom. He took them in tow and they were back at 1.45pm.

Both 1954 and 1955 proved to be quiet years for the lifeboat with only six calls in the two years.

On Friday 12th March 1954 the tanker **Adroity** of London went aground on a newly formed shoal to the north of the Newcome Sand. The *Michael Stephens* put out at 7.20pm in calm weather and the Master said he would not need the lifeboat unless conditions got worse. The lifeboat was back at 9.0pm. At 12.50am (13th) the Coastguard telephoned the Coxswain to say that the tanker was broadcasting a "Mayday" because she was bumping badly and needed help. The *Michael Stephens* launched again at 1.10am and was nearly to the casualty when she floated off the bank. She managed to get to deeper water with the lifeboat standing by and then continued on her voyage in company with another ship of the same Owners. Mr Sydney Taylor the Honorary Secretary suggested, in his report, that the vessel had pumped some cargo overboard (how times change!!).

On Wednesday 6th July 1955 the relief boat *Greater London (CS No.3)* was on exercise with RNLI District Inspector Wheeler on board when a message was received that a man was missing from a capsized yacht **White Lady** off Southwold. Together with the Aldeburgh lifeboat she searched for nearly four hours but no trace was found of the missing man. The other three crew of the yacht had been picked up by the local fishing boat LT352 **Boy Colin**. The lifeboat was back at her moorings at 5.30pm.

On Friday 20th January 1956 the Master of the 16000 ton tanker **British Empress** reported that he was close to **Corton LV** with an injured crewman aboard. The *Michael Stephens* launched at 10.30am with Dr. AC Gee, the Port Medical Officer and Mr William Bullen of the St. John Ambulance

Service on board. The Doctor was put aboard the tanker and eventually the injured man, who had a fractured skull, was transferred, with the Doctor, to the lifeboat. Harry Burgess brought the lifeboat back in a force six WSW wind and choppy sea landing the injured man at 1.15pm. The local newspaper the *Lowestoft Journal* was able to send a photographer on this service and some pictures appeared in the issue for 22nd January. Later the British Tanker Co. Ltd., sent a donation of five guineas to the lifeboat.

At about 3.15am on Wednesday 11th April 1956 the steel dumb-barge **Leeds Saturn** ran aground near the Coastguard lookout having broken from the tug **Dundas Cross** which was towing her, and another barge, from the Humber to Avonmouth. The maroons were fired for the lifeboat and the Lifesaving Apparatus team. The ***Michael Stephens*** launched at 3.30am and found that three men had boarded the barge from the shore. Coxswain Burgess anchored the lifeboat and veered down on the barge and with the help of the three men was able to get a rope secured and tow her off. The barge was towed out to the tug waiting in the North roads. In March 1958 the salvage case was settled in the Admiralty Court which awarded the lifeboat £175 salvage money.

Saturday 1st September 1956 was a grey overcast day with a very strong NE wind and heavy seas. As darkness fell the Coastguard reported flares about a mile south of the harbour. The relief boat ***Greater London (CS No.3)*** launched at 7.45pm and found a vessel had grounded on the Newcome Sand opposite the MAFF Fisheries Laboratory at the old Grand Hotel. As the lifeboat approached, the vessel capsized and sank. Harry Burgess took the lifeboat in amongst the wreckage and found men clinging to oars and floats. He rescued the whole of the crew of nine from the wooden trawler **Les Deux Jeannes** of Boulogne which had been heading for Lowestoft in company with another Boulogne trawler the **En Avant**. The latter boat had seen her sister vessel wrecked but could not get near enough herself. Indeed, having landed the nine men, Harry Burgess took the lifeboat out again to signals from the **En Avant**. Her crew having seen their comrades thrown into the sea had not realised that the lifeboat had rescued them. The Skipper of the **En Avant** followed the lifeboat safely through the banks to the harbour where they were reunited with their fellow fishermen. On the following Monday the shipwrecked mariners were seen off by train to London where they were met by Officials from the French Embassy.

The *Lowestoft Journal* of 7th September reported that William Capps-Jenner, one of the lifeboat crew, had paid tribute to Harry Burgess saying, "It was due to him that we got them all. He just went round them floating in the water with such skill that we crew were just able to pick them out." Lady Somerleyton, President of the Lowestoft RNLI Branch, sent a telegram of congratulation.

At the time of writing (2007) I feel sure, having read all the reports of this service, that to-day it would have been a medal service. Consider the darkness, a gale from the most exposed quarter, men and wreckage floating hither and thither in the water; it took great skill to rescue them all without loss. I know, though, that Harry Burgess was a man of few words and very self effacing. It is quite likely that he saw it as just part of the job.

During the earlier part of 1956 the *Michael Stephens* had been fitted with VHF radio which gave the ability to keep in touch with aircraft and well as the shore.

On Wednesday 24th April 1957 the half-yearly inspection of the lifeboat was carried out by Cmdr. HB Ackworth (Eastern Area Inspector of Lifeboats). The opportunity was taken to carry out an exercise in conjunction with the RAF Air-Sea Rescue helicopter from RAF Horsham St. Faith at Norwich. The exercise was carried out off Pakefield and a member of the helicopter crew was lowered to the *Michael Stephens* at the second attempt. However, when it came to recovering the man, it was found that the hauling winch on the aircraft was jammed. The helicopter landed ashore close to Whapload Road while Harry Burgess took the lifeboat back to her moorings.

The *Lowestoft Journal* of 26th April reported that the aircrewman LAC Blowers said that he had been a lot happier suspended above the lifeboat than rolling back to harbour aboard it. LAC Blowers had been the "victim" during an earlier exercise at Gorleston.

On Friday 2nd August 1957 the Lowestoft Branch of the RNLI held its Annual Lifeboat Day with the relief boat *Greater London (CS No.3)* being open to the public. Inevitably a call came. Just after 4.0pm the lifeboat was off to the aid of two canoeists who had capsized north of the harbour. They were found clinging to the canoe **Roaring Boys** and were returned, with the canoe, to the harbour. Photographs show a good "lifeboat day" crowd on the South Pier to see them come back. The pair were Robert Runacres of Lowestoft and James Harrison of Wokingham, Surrey. Both were from the Army Apprentice School in the latter town. They had been off on a canoeing holiday on the Broads via Yarmouth.

About six weeks later on Saturday 14th September 1957 the longshore boat LT380 **Boy George** developed engine trouble 1½ miles SE of Southwold. *Lloyds List* of 15th September reported that the *Michael Stephens* was launched at 3.0pm and towed the casualty back to harbour. The owner, George Blowers, was alone in the boat when the engine failed and had drifted as far as Kessingland when the lifeboat found him.

A week later another longshore boat LT353 **White Heather** was in trouble while fishing off Kessingland. Just before midnight on Thursday 19th September Bert Beavers, alone in the boat, ran it ashore after the engine failed. At first light the following day he was joined by his brother Jack and the two of them removed all the gear from the boat. At 5.0pm, still being unable to start the engine, they decided to sail the boat to Lowestoft. It was slow progress and by 8.0pm the wind and tide were taking them too close to the Pakefield Church groyne. The alarm was raised from the shore and the maroons were fired for the lifeboat and the Coastguard Lifesaving Apparatus team. The *Michael Stephens* was away quickly but Harry Burgess found that the casualty was too close ashore for him to get alongside. A large crowd of people on Pakefield beach and cliffs watched as the LSA team fired a line across the boat, now anchored a few yards offshore. A rope was attached to the longshore and a band of willing helpers pulled the **White Heather** and her crew of two safely ashore.

The *Lowestoft Journal* of 25th October 1957 carried extensive reports of the Trafalgar Day Review of the Herring Fleet. This was held off Great Yarmouth during a week that also saw the Annual Herring Banquet at Yarmouth and the opening of the International Fisheries Exhibition at Lowestoft.
The *Michael Stephens*, dressed overall, left bright and early at the head of a good number of drifters from the port. The paper reported that, "Thousands lined the sea front to see drifters and longshore boats, dressed overall, pass between the Britannia Pier and **HMS Wave** from which the Commander-in-Chief Nore, Admiral Sir Frederick Parham took the salute. Ships of the Royal Navy provided a background to the event together with the Polish training ship **Turletski**. Special guests were ferried from South Quay in Yarmouth out to **HMS Wave** by **HMS Squirrel**."
The paper also reported that a display of fish from Lowestoft Fish Market was staged at the South Pier Pavilion in connection with the Fisheries Exhibition.

A sad event took place on Sunday 19th January 1958 when a boy was washed off the sea wall and lost. Two 13-year olds were walking along the sea wall when they were hit by an exceptionally large wave out of the many that were hitting the wall. Gary Aguss managed to hang on but James Balls was swept off the wall. The Coastguards were summoned and the *Michael Stephens* launched at 8.57am. Coastguard Cecil Scott was off duty but was quickly on the scene and, donning his lifejacket and lifeline, which was held by people on the wall, he went into the sea but to no avail. The lifeboat searched close inshore helped by a helicopter and RAF Rescue Launch for nearly two hours but nothing was found. James was the son of Sub-Officer James Balls of the Lowestoft Fire Brigade. The *Lowestoft Journal* of 24th January reported great praise from onlookers for Cecil Scott, with Mr Brian Balls of Battery Green Terrace saying, "It was really hopeless, and the man was in great danger of being dashed against the wall."

The station was quiet for a while until Bank Holiday Monday 4th August when a capsized canoe with a person in the water was reported about two miles south of the Coastguard lookout. Thousands of holidaymakers on the sea front saw the *Michael Stephens* leave soon after 11.0am together with a RAF helicopter overhead. In the end it turned out to be a 'false alarm with good intent'. The canoeist, Colin Evans, had left his canoe for a swim (it must have been a calm day) and a person ashore seeing the empty canoe and a person in the water assumed the worst.

On the following Monday, the 11th, a converted ship's lifeboat, the **Knud**, ran shore on the South beach. The *Michael Stephens* was launched at 1.15pm and the Coastguard Lifesaving Apparatus team also attended. The owner and crew of the **Knud** was Mr Geoffrey Seago and he was pulled ashore by Coastguard Ernest Tucker. Other Coastguards and onlookers, together with Mr Seago, pulled the boat clear of the sea but eventually she broke up. The lifeboat was unable to offer any help owing to the casualty being on the beach and she returned to station.

On Sunday 5th October 1958 the Yarmouth longshore boat YH243 **Harnser** got her nets tangled round the propeller in deteriorating weather. Her two crew anchored off Corton and burnt flares and signalled for help. These were seen by Inspector R Creighton of Yarmouth Police and he alerted the Coastguard. The relief boat *Cunard* was launched under Coxswain Harry Burgess at 1.10pm since the Gorleston lifeboat was already out to the ring-netter YH61 **Cossack**. The *Cunard* got a tow rope onto **Harnser** and, despite it parting twice, got her safely into Gorleston. The lifeboat was back at Lowestoft by 4.30pm.

On Sunday 1st February 1959 the longshore boat LT374 **Marina** developed engine trouble off Pakefield. The two men on board, Brian Balls and Henry Wymer, burnt petrol-soaked rags to attract attention. The *Michael Stephens* launched at 5.25pm into a fresh NE wind and choppy sea. The boat was taken in tow and both were back by 5.40pm. By coincidence, Edward Balls a member of the lifeboat crew was the uncle of Brian Balls. The *Eastern Daily Press* for the next day reported Brian saying, "The trip wasn't wasted because we caught plenty of whiting and codling."

Later in the year fire at sea developed into a major event for onlookers on the shore. On Saturday 27th June 1959 the mv **Staniel** of Cardiff was southbound off Aldeburgh when it was discovered that the deck cargo of hay was on fire. The Master, Captain H Slugar, turned his vessel round to run before the wind. Both the Aldeburgh and Lowestoft lifeboats launched and the *Michael Stephens* took off most of the crew of the **Staniel** a mile or so east of Sizewell Bank and then escorted the vessel towards Lowestoft. She also put six lifeboatmen on board to help fight the fire. However the fire spread rapidly and it was decided to beach the vessel at Pakefield. The remaining crew were taken off by breeches buoy and some Lowestoft Firemen went on board by the same means but eventually had to leave because of the danger of exploding fuel tanks.

At the end of 1959, on Monday 7th December, the mv **National Fighter** of Monrovia was anchored three miles SE of Lowestoft, close to the Newcome Sand, with engine trouble. Her Master fired distress rockets when one of his anchors parted in the SE gale and rough seas. The *Michael Stephens* left at 7.45am and the Master asked Coxswain Burgess to stand by while he steamed ahead to take the strain off the remaining anchor. However that too parted. The Master then found that, despite steering problems, he could make headway by steaming astern to the east. The lifeboat stood by until he was five miles out and clear of danger. Eventually the **National Fighter** was towed to Gravesend by four tugs – **Jean Bart**, **Vanquisher**, **Cervia** and **Gondia**. The *Lowestoft Journal* for 11th December reported that, for the first time since the war, the Kessingland Rocket Apparatus team were called out in case the casualty drifted ashore there. In the event they were not needed.

2.2 - The Lifeboat in the 1960's

It turned out that 1960 was a quiet year for the lifeboat with only six calls the whole year.

In June 1960 the RNLI Journal *The Lifeboat* reported the retirement of the Honorary Secretary Mr Sydney Taylor OBE. Mr Taylor had served as Secretary for 35 years and was awarded the Thanks of the Institution on Vellum. Mr Taylor was succeeded by Mr EK Tucker.

In August 1960 the relief boat **Cunard** took part in an exercise with the Fire Service and the RAF. HM Inspector of Fire Services, Mr SH Charters and the Suffolk and Ipswich Chief Fire Officer Mr HF Griffiths were on board as were some men from the Lowestoft brigade. The firemen demonstrated the use of water and foam to fight fires at sea. Later, the RAF helicopter showed how to get men and equipment on board a vessel.

On Tuesday 1st November the Coastguard saw red flares ½ a mile SSW of the lookout at about 1.0am. Harry Burgess took the *Michael Stephens* out and found a small longshore boat LT84 **Maggie Jane** with a crew of three. They had set off the previous afternoon and had anchored about 300 yards off the Claremont Pier. When they came to start the engine for the return it would not fire and after two hours of trying during which time they had started dragging towards the pier, they decided to signal for help. The lifeboat was able to tow the boat, which was owned by Mr Clarkson a local dentist, to the harbour just after 2.0am.

On Monday 5th December 1960 the Polish trawler SZN.262 **Pradnik** reported that she had an injured man on board and needed medical help. The *Michael Stephens* put out at 9.0am with Dr. AC Gee on board and came up with the trawler about four miles east of Southwold. The Doctor was put aboard and the injured engineer, who had been badly scalded by a burst steam pipe, was treated. It was decided that the trawler would proceed to Yarmouth where the injured man would be landed to hospital.

A much busier year followed. On Tuesday 7th February 1961 it was reported that the Boulogne trawler **Georges Ferges** had a sick man on board and would arrive off Lowestoft at 4.0am for help. The *Michael Stephens* launched with a Doctor and a Pilot on board into a WSW gale and very rough sea. The Pilot was taken in case the sick man could not be transferred to the lifeboat and the trawler would have to be piloted into harbour. In the event, the Doctor reported that the man was dead and could not have been saved since he had a severely fractured skull. The trawler Skipper said he would return to France with his dead colleague and the lifeboat was back at 7.20am.

Two weeks later on Friday 24th February 1961 the *Michael Stephens* took part in a service to a vessel on fire. The Norwegian mv **Gudveig** reported an engine room fire and was a mile east of the NE Cross Sand buoy in thick fog. The Gorleston lifeboat *Louise Stephens* (ON820) put out and the *Michael Stephens* launched later at 5.40am with an Officer and men of the East Suffolk Fire Service. She reached the casualty about 8.30am and put the firemen and their equipment on board. Twenty-eight men from the **Gudveig** had left in a ship's boat and were picked up by the mv **Tennyson** which was standing by. They were transferred to the Gorleston lifeboat which landed them at Yarmouth. Later, the Caister lifeboat *Jose Neville* (ON834) went out with petrol for the fire pumps. At 4.0pm both Caister and Lowestoft lifeboats left the scene to search for a small boat with a press photographer on board which was reported missing. Fortunately the boat and two people had been picked up by a passing ship and so both lifeboats were able to return to their respective stations. Meanwhile the crew were put back aboard the **Gudveig** and the Gorleston lifeboat took off the firemen and their equipment landing them at Yarmouth. The fire had been extinguished by 5.0pm and a Dutch tug the **Gele Zee** towed the casualty to Immingham where she had been bound from Casablanca with phosphates.

This service generated a lot of press coverage, not least because of the involvement of the Fire Service. The Lowestoft Honorary Secretary reported that this "job" had proved the value of exercises his station boat had held with the Fire Service over the past year. He could not fault the work by the *Michael Stephens* which had been involved because the Gorleston boat had not held any exercises with the Fire Service up till then. At the time, the Fire Service Divisional Officer Mr RH Bishop praised Coxswain Harry Burgess for the "uncanny" way he found the ship in dense fog. The Coxswain thought it the longest service – 13 hours – that he could remember.

On 28th March 1961 *Lloyds List* reported the **Gudveig** sold for breaking up for £39000, she had originally been built in Gothenburg in 1932. Later the *Eastern Daily Press* of 20th April reported that a ship's boat from the **Gudveig**, towed in by Gorleston lifeboat at the time, had been on Gorleston quay since then. The Martham Boatbuilding Co had bought it from the owners and were to use it as a work boat in their yard.

On Tuesday 16th May 1961 the German mv **Ameland** of Gluckstadt grounded on the Holm Sand. The relief boat ***Cecil & Lilian Philpott*** put out at 4.37am and found the vessel with a heavy list to starboard. Since she had a deck cargo of timber there was the danger that this might slip and cause the casualty to capsize. The lifeboat stood by through low water until the tide made enough to float the vessel. The lifeboat then escorted her into Lowestoft. Fortunately for everyone the sea was calm and the weather kind.

On the 28th June the relief boat was taken up to Leo Robinson's yard at Oulton Broad for refit and the station boat *Michael Stephens* brought back by Coxswain Burgess and five of his crew. I have the receipts for a 5s.0d transit of Mutford Lock between Lake Lothing and Oulton Broad. A note from Harry Burgess confirms the crew as himself, J Saunders, V Tuck, T Balls, W Capps-Jenner and J Polkinghorn the mechanic.

On Sunday 2nd July 1961 a 14ft half-deck yacht the **Sujan** was launched off Corton beach with a crew of two men and a woman. A few minutes later in choppy seas produced by the ebb tide being against the wind she capsized. All three people clung to the upturned hull and were kept under observation by the Duty Coastguard at Gorleston, Mr Arthur Weston. The *Michael Stephens* was launched at 4.15pm and a RAF helicopter was scrambled. Owing to the sea state they had some difficulty in locating the casualty but since the Coastguard could see the yacht he was able to direct both units. The lifeboat rescued the two men while the helicopter took on board the woman. The yacht was righted and towed to Gorleston by the *Michael Stephens* who landed the two men there. The *Lowestoft Journal* of 7th July reported that thousands of holidaymakers had a grandstand view of this combined rescue operation by boat and helicopter.

On the evening of Monday 21st August 1961 two youth leaders from the London Boys Camp at Kessingland put off from the beach in a canoe. Sometime later they were seen to have capsized. The ***Michael Stephens*** was called at 6.0pm. At the same time four men put off in an 18 foot dinghy from Kessingland. They were able to rescue the two men and get back to the beach but not without shipping a good deal of water. A helicopter was also involved but the lifeboat was recalled. The *Lowestoft Journal* of 25th August gave some details. The two rescued were Mr Frank Castella and Mr Tom Hibbert both of London. They were saved by Mr Cyril Lyne, Mr Derek Lyne, Mr Michael Lyne and Mr Arthur Veness. Mr Castella, a Spaniard said, "Everything was fine, we had a sail up but then a gust of wind hit us and over we went." Both Londoners were taken to the home of Mrs P Edmonds who had originally alerted the Coastguard. Two local boys were also involved; Terry Dowding and Bruce McMeekin both swam out but were able to return when they saw the rescue had been carried out.

On Monday 22nd January 1962 the wooden longshore boat LT654 **Dulcie Doris** was fishing some miles off Lowestoft when the engine failed. The Lowestoft trawler LT270 **Unda** took off the two crew and took the longshore in tow. Alfred Buck, Skipper of **Unda**, asked through Humber Radio, for the lifeboat to go out and take the men from him. ***Michael Stephens*** launched at 2.40pm into a strong WNW wind and a nasty swell. She met the fishing boats two miles outside the **Corton LV** when the tow had just parted. Two of the lifeboat crew were put aboard the casualty to remake the tow which the lifeboat took over from **Unda**. The two crew from **Dulcie Doris** were taken onto the lifeboat and the tow got underway. The rough conditions may just have been too much for the 60-year old longshore and when they were just off the SE Corton buoy she took a nosedive and sank. The original problem had been a broken prop shaft. Her two crew were Mr Ken Stephenson and Mr Leslie Upson who had gone out to recover lines they had laid the previous Saturday. The owner of the longshore was Mr HG Case of Wrentham who sent a £10 donation to the lifeboat.

On Monday 12th February 1962 the Hartlepool trawler HL48 **John O'Heugh** rescued 18 men from the burning mv **Fountains Abbey** 60 miles NE of Cromer. She headed for Lowestoft with the rescued, some of whom were injured. At 8.0am on the following day the ***Michael Stephens*** launched with a Doctor on board to meet the trawler at the North Corton buoy. About an hour later Dr. Watson was put aboard and the lifeboat followed the trawler to Lowestoft, where they arrived about 10.0am. The injured men were taken by ambulances to hospital.

The **Fountains Abbey** had been built in 1954 by Hall, Russell at Aberdeen for Associated Humber Lines. The vessel was totally burnt out and was towed to Ijmuiden, Holland by three tugs – **Titan**, **Simson** and **Stentor**.

On Saturday 21st July 1962 the trimaran **Nimble Venture**, which had been built in Yarmouth, was on her delivery voyage to Ipswich. In a stiff SW breeze off Sizewell she lost part of her mast and the crew decided to put back. When they were near Lowestoft they put up flares for assistance as the strong ebb tide was rapidly carrying them past the harbour mouth. The *Michael Stephens* launched just after 4.0pm but found on arrival at the casualty, that two of the Royal Norfolk & Suffolk Yacht Club launches were towing the trimaran. In the force seven SW wind the lifeboat escorted them towards the harbour. As they were lining up for the harbour entrance the tow broke and the trimaran was swept north. The lifeboat managed to get the tow remade and got her in safely at 6.30pm. The owner of the trimaran, Sqn.Ldr. N Clarke, sent £10 for the RNLI.

Two days later a very sad event happened. Just after breakfast on Monday 23rd July two young boys drifted away from Pakefield beach on an air bed. The Coastguard reported that they had been seen to fall into the sea and the *Michael Stephens* launched urgently just after 9.0am. Together with a helicopter and some local longshore boats she searched for over three hours to no avail. Conditions were good and the sea calm but neither of the boys was found.

Late on the night of Thursday 9th August 1962 the French trawler, B.2542 **Pierre Louis**, with some gear round her propeller tried to enter the harbour. She could only make slow speed and she was carried onto the North extension. The *Michael Stephens* launched at 11.55pm and the Rocket Brigade was also called. Two Coastguard Officers scrambled along the extension and got on board the vessel. With their help the lifeboat was able to get a tow rope fixed by which means the casualty was towed off with the help of the French Skipper going full astern on the engine. The two Coastguards – Glenn Clarke and David Strathearn – played a very important role in this rescue since none of the Frenchmen could understand English.

Saturday 10th November 1962 was a day of strong ENE winds and rough seas. Some miles out the Scottish drifter BF372 **Silver Wave** developed engine trouble. She was taken in tow by the Yarmouth tug **Richard Lee Barber**, escorted by the drifter FR185 **Bdellium**. As they were approaching

Lowestoft the tow parted and **Silver Wave** was left drifting towards the sea defences north of the harbour. She made distress calls on her siren but the Skipper of the **Bdellium** was quick. He managed to get alongside and get a rope aboard his comrade by which means he got her into the harbour. The *Michael Stephens* had launched but was only needed to stand by.

The *Lowestoft Journal* of 16th November reported that Skipper Frank West of the **Bdellium** displayed remarkable seamanship in getting alongside the **Silver Wave** in very dangerous conditions. Skipper George West (no relation) said, "It was a very fine piece of seamanship to get alongside and get us under tow again."

Over the weekend of 19th/20th January 1963 the South Holm buoy broke adrift from its moorings and ended up on Pakefield beach. Predictably, late on the 21st, the mv **Hoocreek**, on passage to Queenborough and not being aware of the missing buoy, ran aground on the Newcome Sand. For an hour or so the Coastguards signalled her with a morse lamp but no reply was received. At 2.30am on the 22nd the relief boat *Cunard* put out and found that the **Hoocreek** had just refloated. The Master asked Coxswain Burgess to escort him to deeper water. The *Eastern Evening News* of 22nd January reported a coincidence for Tommy Knott the mechanic of the lifeboat. In 1962 Tommy had been a passenger on a ferry entering Dover harbour when he reported seeing flares offshore. He had been quick off the ferry and had been able to go out with the Dover lifeboat to find mv **Hoocreek** abandoned after making water.

On Sunday 14th April 1963 the Aldeburgh relief lifeboat *George & Elizabeth Gow* (ON827) developed engine trouble. The *Michael Stephens* went down to Aldeburgh to tow her back to Lowestoft for repair. On Sunday 12th May, after being repaired, she was on passage back to Aldeburgh when she broke down again. Sails were set and she headed back for Lowestoft to be met by the *Michael Stephens* off Southwold. She was towed back to Oulton Broad for further repairs.

On Sunday 30th June 1963 the *Michael Stephens* carried out her final duty as the Lowestoft station lifeboat. She launched at 5.0pm and headed south to meet the new station boat *Frederick Edward Crick*. When both boats were off the Claremont Pier the Coastguard fired the maroons and a large crowd turned out on the piers and docks to welcome the new boat. Later, a number of invited guests including Borough Council members, Police

Officers, the Lifeboat Committee and Ladies Guild members went for a trip on the new boat.

Harry Burgess, Frank Swan, Jack Rose and Tommy Knott from the Lowestoft crew together with Mr GR Walton (RNLI Assistant District Inspector) and Mr B Rickard (RNLI District Engineer) formed the crew who brought the new boat back from Cowes, Isle of Wight. They had left in the early hours of Saturday, stopped overnight at Ramsgate, and left at 9.40am on the Sunday for Lowestoft.

During her time at Lowestoft the **Michael Stephens** launched 134 times and saved 73 people.

The new boat had two calls within a month of arriving at Lowestoft. On Sunday 28th July 1963 the **Frederick Edward Crick** launched at 5.0pm to the aid of a boy swimming off Pakefield. He had tried to swim out to a yacht race marker buoy but got severe cramp. Fortunately for David Evans of London, who was staying at the Pakefield Holiday Camp, he was seen by the crew of the yacht **Bittern** who were able to fish him out. The **Bittern** was owned by Peter Catchpole, who, with Richard Hailey and Paul Long was taking part in a race from Southwold to Lowestoft. They shipped a lot of water in the rescue operation so the lifeboat gave them a tow back to Lowestoft.

The following day at 2.0pm came the second call. The small fishing boat HL135 **Dolly Graham** was on passage from Scarborough to Newhaven when she grounded on Corton Sand. The *Frederick Edward Crick* found her in a difficult position and when she went alongside to pass the tow rope the lifeboat was in only four feet of water. After two breakages the tow succeeded and the fishing boat came clear with a broken rudder. She was towed to Hamilton Dock with her crew of two. The *Eastern Evening News* of 29th July reported that the vessel took a lot of water when she heeled over on hitting the sand. Her owner was Polish, Mr Joe Sitco, he had bought the vessel and was sailing her back to Newhaven where he was a longshore fisherman.

On Saturday 10th August 1963 the Coastguard reported that the Norwegian mv **Skaansund** was towing a yacht, **The Fly**, with three men on board. They were three miles SE of Southwold and asked for the lifeboat to take over the tow. The **Frederick Edward Crick** launched at 11.50am but soon after it was learned that the Southwold Harbour Master, Mr Upcraft, in his launch **Silver Surf** had taken the tow. The lifeboat escorted both to Lowestoft.

On Saturday 9th May 1964 the motor cruiser **Boy Leslie**, on passage from Yarmouth to Lowestoft, lost power and was swept onto the Newcome Sand. She was bumping badly and sprang a leak. The crew of three fired distress flares and the *Frederick Edward Crick* was smartly away at 5.30pm. The three people (a man and woman and a girl) were taken into the lifeboat and replaced by three lifeboatmen to bale the boat out. The cruiser was lashed alongside the lifeboat for the journey back to harbour. Harry Burgess recorded, "We just got round the dump head when she sank but by going full ahead we managed to run her on a mud bank." On the following day the tug **Lound** salvaged the cruiser.

On Monday 26th October 1964 the Coastguard reported a small boat sinking off Pakefield beach. The *Frederick Edward Crick* put out at 7.10pm and Coxswain Burgess recorded in his Log Book the following. "Using our searchlight and guided by shouts from the beach we found Paddy Mewse in the longshore boat LT283 **Boy Roy** the in breakers, sinking from the weight of herring he had caught. We got helpers on the beach to pull his nets ashore with the fish, used a small boat to connect our tow line and towed him back to harbour."

At the end of 1964 Mr EK Tucker relinquished the post of Honorary Secretary and was replaced by Dr. Bruce Markham a local GP.

The *Frederick Edward Crick* was out for over 12 hours on Saturday 20th February 1965 standing by the stern trawler SN58 **Sailfin** which ran aground off the South Pier, while trying to enter the harbour in an easterly gale. Harry Burgess reported that after four attempts he managed to get a line on board the casualty but she was too heavy for the lifeboat to move her. So he took a line to the tug **Richard Lee Barber** of Yarmouth and at high water she managed to tow the trawler off stern first. In the North Roads they moved the tow to the bow and headed for the harbour. Coxswain Burgess went on, "They got too close to the North Pier and the trawler grounded on the shingle bank alongside the North Extension. The Coastguard Lifesaving team took the crew off leaving just the Skipper and an engineer."
By now it was low water and the tug and lifeboat waited in the harbour until 8.0pm when they went out again. The trawler was safely in by 11.0pm.

In July the Admiralty Court settled a salvage claim and awarded the lifeboat £53.

On Wednesday 18th August 1965 a holidaymaker reported seeing flares three or four miles out off Corton. The relief boat *Elizabeth Elson* was launched at 2.40pm and searched for some time but found nothing. Harry Burgess thought it was probably the sun reflecting from the wheelhouse windows of a passing ship which the visitor mistook for flares.

A week later, on 25th August, there was another fruitless call. The German vessel mv **Tannenberg** reported a serious list of 35° after her deck cargo of timber had shifted, 24 miles SE of Lowestoft. Since the Aldeburgh lifeboat was already out on a call the relief boat *Elizabeth Elson* launched at 1.50pm into a westerly gale. After about half an hour a message was received that the mv **Winchester Brook** was standing by the German who was under way to pick up a pilot at the **Sunk LV**. A tug was also heading for the scene and so the lifeboat was recalled.

During the morning of Saturday 30th October 1965 Harry Burgess was in the Coastguard lookout watching the trawler LT245 **JAP** which was aground just outside the harbour. There was a SW wind gusting to force seven creating a nasty broken sea. The trawler was rolling and bumping badly and soon hoisted a distress signal. The *Frederick Edward Crick* was away at 9.55am and in the meantime the drifter LT382 **Wisemans**, of the same owners, had got a line on board the **JAP** but had herself gone aground. The **Wisemans**, a small wooden drifter did refloat and got into the harbour. Harry Burgess reported that the lifeboat stood by for 45 minutes until the **JAP** came off on the rising tide. They then escorted her into the harbour.

On Tuesday 11th January 1966 the trawler LT402 **St. Georges** had just put to sea when she hit the bottom off Corton and began to leak. The Coastguard reported that she was burning red flares just inside the West Corton buoy. The *Frederick Edward Crick* launched at 3.30pm and the Lifesaving Apparatus team were put on standby. Water had got into the trawler's fuel tank and the engine stopped. She did manage to get an anchor down in the force seven easterly wind. They had no power or lights and her skipper asked the lifeboat to stand by in case the anchor parted. The lifeboat also relayed messages ashore until the trawler managed to get power restored. The lifeboat stood down at 10.0pm after the trawler's owners sent out their vessel LT340 **Silverfish** to stand by. The lifeboat was able to escort

St. Georges into the harbour the following morning after the trawler's engine had been restarted during the night.

A more unusual call came on Sunday 27th March 1966. The drilling rig **Constellation** was about 40 miles east of Lowestoft in a force 10 westerly storm. The tug towing her was unable to prevent the rig from driving downwind and the lifeboat was requested in case the rig met with an accident. The *Frederick Edward Crick* launched at 8.10pm and headed east. At 11.0pm they received a call that conditions at the rig had improved and their help was not needed. The lifeboat turned and headed for home, which must have been a very uncomfortable passage. The rig reported at times that she was pitching up to $12\frac{1}{2}°$ and if it had got to $15°$ the drilling derrick would have collapsed. Harry Burgess and his crew were back at 2.45am (28th).

On Sunday 10th April 1966 Harry Burgess was informed that the trawler LT88 **Ormesby Queen** was heading back to port with a sick man on board. The lifeboat crew stood by and at 10.40pm, nearly two hours later than the planned arrival time, the trawler reported that she was anchored south-west of the North Corton buoy. She would not risk coming closer because of the low tide and thick fog. The *Frederick Edward Crick* launched at 10.45pm and searched for over an hour without success. Eventually the trawler was asked to sound her siren and this helped to locate the vessel. The sick man was taken on board the lifeboat which landed him ashore to an ambulance at about 2.0am (11th). The Station Honorary Secretary, Dr Markham, noted in his report, "….radar would have saved a lot of time."

A larger ship than those usually attended to by the Lowestoft lifeboat was the subject of a service on Sunday 3rd July 1966. The British India Steam Navigation Co. liner ss **Dunera** was on a schools educational cruise. The Master reported that he wanted to land a sick schoolgirl with some urgency. The *Frederick Edward Crick* launched and met the liner near the South Holm buoy. The girl was put in one of the ship's lifeboats which was lowered to the water with the lifeboat going alongside to take her on board. She was attended to by Dr Watson on the passage back and landed to an ambulance for hospital suffering from appendicitis.

Twice in six days an embarrassing incident happened to a small fishing vessel LT448 **Seafarer** owned by Jack Hales. On Saturday 15th October 1966 he burnt flares just outside the harbour. The *Frederick Edward Crick* was out at 1.0pm in a sharp southerly breeze and choppy sea. The fishing boat had been out testing some gear for the MAFF, Fisheries Laboratory just off the port and had got a rope round the propeller. The lifeboat towed them in, the five crew included two scientists from the Laboratory.

On the following Thursday, the 20th, the same thing happened again about two miles south of the harbour and again the lifeboat towed them in with some more gear wrapped round the propeller.

The author remembers this incident well since, at that time, he was employed at the Laboratory. He recalls much hilarity amongst his colleagues who were not involved; the two involved were only too grateful that the lifeboat got to them in time.

On the morning of Thursday 15th December 1966 the small tug **Kendal** of Rochester was towing the small trawler LT358 **Garibaldi J** from Rotterdam to Lowestoft. The tug ran onto the Newcome Sand and the trawler ran into the stern of the tug smashing her stem in the process. To save further damage the trawler was cast adrift. Harry Burgess launched the *Frederick Edward Crick* at 10.55am and picked the trawler up almost on the South beach and towed her back to harbour. The lifeboat then went back and stood by the tug. By 1.0pm the tide had fallen and the swell was breaking on the bank. The tug Skipper asked the lifeboat to take off his crew. Harry Burgess records that they went alongside and with the echo-sounder showing zero water took off eight men, sustaining some damage in the process. Later in the evening the pilot cutter took out the crew to their vessel and as the tide rose she came off and they were able to get her into the harbour.

On Saturday 14th January 1967 the relief boat *Elizabeth Elson* launched to the fishing boat FR240 **Mizpah** with a crew of two, which was aground just outside the harbour. The lifeboat went out at 4.50pm and towed the boat back by 6.0pm. At the time of this service the lifeboat crew and their wives were running the annual children's party at the Lifeboat Social Club. The Mayor of Lowestoft, Mr Roy Burgess, was attending. The *Eastern Evening News* of 16th January reports that the Mayor donned oilskins and went out with the lifeboat. It was believed to be the first time a Mayor of Lowestoft had gone to sea on a lifeboat service. Mr Burgess described it as, "A terrific experience."

On Saturday 13th May the Coastguard reported that a small yacht, the **Contango** with a crew of six, was aground on the Newcome Sand and was flying a distress signal. Harry Burgess was away with the *Frederick Edward Crick* at 8.25pm and found the yacht with her engine broken down and a strong flood tide driving her onto the bank. The lifeboat anchored, veered down in just enough water and passed a tow line. The yacht and crew (one a woman) was towed in by 9.0pm.

On Wednesday 5th July 1967 the new Cromer lifeboat *Ruby & Arthur Reed* (ON 990) launched on her first service at that station to a small trawler BM74 **Renovate** which was broken down off the north Norfolk coast. There were several other vessels helping including the Fishery Protection vessel **HMS Belton**, the mv **Fiducia II** of Holland, the Happisburgh inshore lifeboat, and the British mv **Ecctonia**. Eventually this last vessel took the casualty in tow and brought her to Lowestoft Roads. The *Frederick Edward Crick*, under Billy Thorpe the Second Coxswain, met them at the West Holm buoy and took over the tow. The fishing vessel was safely brought into Hamilton Dock at 7.15am.

All too often, around our coasts, lifeboats are involved in services which have a sad end. On Thursday 13th July 1967 the *Frederick Edward Crick* was launched at 3.0pm to search for a young swimmer swept out to sea. The lifeboat was joined by the beach rescue craft, two helicopters, and the fishing boat LT451 **Tearaway**. They all searched for over four hours but no trace of the swimmer was found.

Later in 1967 on Wednesday 1st November, the Coastguard reported that the four ton sloop **Vale** was on the Newcome Sand bumping and rolling in the southerly gale. The *Frederick Edward Crick* launched at 2.35pm but could not get nearer than 50 yards since the casualty was on the widest part of the shoal. The lifeboat crew fired lines and at the third try got one on board the vessel. By going slowly astern they were able to get the casualty off and finally into the yacht basin. The sloop's crew were Mr Gilbert Aitken, Managing Director of Colman's Mustard and his son, both of Norwich and Mr Terrence Poole. They had been on passage from Orford to Lowestoft.

1955	*Michael Stephens*	
21 Jan	trawler, **Grasby** (LT267), Newcome	stood by
23 Jun	m cruiser, **Marina**, Pakefield	saved 2 + boat
	Greater London (CS No.3)	
6 Jul	yacht, **White Lady** capsized off Southwold	assisted
1956	*Michael Stephens*	
20 Jan	tanker, **British Empress**, injured man	landed 1
11 Apr	dumb barge, **Leeds Saturn**, N beach	towed off
12 Jun	fb, **Eva May** (LT266), overdue	saved 1 + boat
	Greater London (CS No.3)	
19 Aug	**Vivo** - no details	ns
1 Sep	trawler, **Les Deux Jeannes** of Belgium	saved 9
1 Sep	trawler, **En Avant** (B.2665) of Belgium	escorted in
	Michael Stephens	
14 Sep	fb, **Eva May** (LT266), Benacre	ns
1957	*Michael Stephens*	
28 Jun	fb, **Pedro** (LT289), Kessingland	saved 2 + boat
	Greater London (CS No.3)	
20 Jul	flares	ns
2 Aug	canoe, **Roaring Boys**, N roads	saved 2 + canoe
22 Aug	swimmer, Hopton	ns
24 Aug	yacht, **Garlock** capsized	saved boat
27 Aug	fb, **Dot** (YH386), Pakefield	saved 1 + boat
27 Aug	swimmers off Southwold	ns
	Michael Stephens	
14 Sep	fb, **Boy George** (LT380), Walberswick	saved 1 + boat
20 Sep	fb, **White Heather** (LT353), Pakefield	ns
21 Oct	took part in review of Herring Fleet	

1958	*Michael Stephens*	
5 Jan	flares	ns
19 Jan	boy washed off sea wall	ns
4 Aug	canoe capsized, S roads	ns
11 Aug	yacht, **Knud**, N beach	ns
2 Sep	swimmer	ns
	Cunard	
5 Oct	fb, **Harnser** (YH243), Corton	saved 2 + boat
1959	*Michael Stephens*	
7 Jan	mv **Borthwick** of Glasgow, Corton	stood by
1 Feb	fb, **Marina** (LT374), Pakefield	saved 2 + boat
17 Feb	mv **Seaford** of London	ns
22 Mar	wreckage of yacht **Daisy**	ns
27 Jun	mv **Staniel** of Cardiff, on fire	saved 7 + 1 bird
12 Jul	dinghy capsized	ns
	Cunard	
6 Sep	fb, **Young Paul** (LT120), N roads	saved 2 + boat
	Michael Stephens	
7 Dec	ss **National Fighter** of Monrovia, Newcome	stood by
1960	*Michael Stephens*	
26 Jun	canoe capsized, Southwold	ns
	Cunard	
14 Aug	small boat capsized, Covehithe	ns
	Michael Stephens	
22 Oct	fb, **Kathleen Ann** (YH260)	saved 2 + boat
1 Nov	fb, **Maggie Jane** (LT84), S roads	saved 3 + boat
5 Dec	trawler, **Pradnik** (SZN.262) of Gdynia	took out doctor
18 Dec	fb, **Tempo** (YH327), N roads	saved 3 + boat

1961   *Michael Stephens*

6 Feb	mv **Henfield**	ns
7 Feb	trawler, **Georges Ferges** of Boulogne	took out doctor
24 Feb	mv **Gudveig** of Oslo, on fire, Cross Sand	took out firemen

Cecil & Lilian Philpott

5 May	parachute into the sea	ns
16 May	mv **Ameland** of Gluckstadt, Holm	assisted
23 May	dinghy capsized, Southwold	ns

Michael Stephens

2 Jul	dinghy, **Sujan** capsized	saved 2 + boat
31 Jul	m cruiser, **Careema**	saved 2 + boat
6 Aug	yacht, **Kapella**, Southwold	ns
9 Aug	dinghy, **Pook**, Walberswick	ns
21 Aug	canoe capsized, Kessingland	ns
23 Oct	drifter, **Henrietta Spashett** (LT82)	assembly
25 Oct	fb, **Brenjean** (LT86), S roads	saved 2 + boat

1962   *Michael Stephens*

22 Jan	fb, **Dulcie Doris** (LT654)	landed 2
6 Feb	mv **Sommen** of Sweden, sick man	landed 1
13 Feb	trawler, **John O'Heugh** (HL48)	took out doctor
2 Jun	tanker, **Esso Stockholm** of London, sick man	landed 1
1 Jul	trawler, **Vestrouwen** of Nieuport	ns
21 Jul	trimaran, **Nimble Venture** of Ipswich	saved 2 + boat
23 Jul	swimmers, Pakefield	ns
9 Aug	trawler, **Pierre Louis** (B.2542), N extension	saved 14 + boat
3 Sep	yacht capsized, Southwold	ns
13 Sep	swimmer, S beach	ns
16 Sep	dinghies adrift, Southwold	ns
6 Nov	tanker, **Esso Parentis** of Le Havre, sick man	landed 1
10 Nov	drifter, **Silver Wave** (BF372)	escorted

1963   *Cunard*

22 Jan   mv **Hoocreek** of London, Newcome   escorted

                Michael Stephens

14 Apr   ***RNLB George & Elizabeth Gow***   towed in
12 May   ***RNLB George & Elizabeth Gow***   towed in
14 May   dinghy, Corton   ns
29 May   flares   ns
12 Jun   trawler, **Moreleigh** (LT170), N extension   ns

                Frederick Edward Crick

28 Jul   swimmer, Pakefield   ns
29 Jul   fb, **Dolly Graham** (HL135), Corton   saved 2 + boat
10 Aug   yacht, **The Fly** and mb, **Silver Surf**   escorted
11 Oct   yacht, **Leila**, Holm   saved 2 + boat

1964   *Frederick Edward Crick*

11 Jan   mv **Rose of Lancaster** of London, sick man   landed 1
19 Mar   trawler, **Tobago** (LT182), N extension   ns
 9 May   m cruiser, **Boy Leslie**, Newcome   saved 3 + boat
10 May   flares   ns
14 Jun   canoes, S beach   ns
24 Aug   m cruiser, **Osprey**   saved 4 + boat
10 Sep   mv **Filius** of Germany   ns

                Elizabeth Elson

 1 Oct   yacht capsized, Corton   ns

                Frederick Edward Crick

26 Oct   fb, **Boy Roy** (LT283), Pakefield   saved 1 + boat
31 Oct   flares   ns

1965   *Frederick Edward Crick*

20 Feb   trawler, **Sailfin** (SN58), South Bank   stood by
20 Feb   trawler, **Sailfin** (SN58), N extension   stood by
30 Mar   fishing boats in a gale   stood by
 7 Jun   tanker, **Olympic Hill** of Monrovia, sick man   landed 1
12 Jun   m cruiser, **Fortuna II**, S roads   saved 3 + boat

1965 (cont)

Elizabeth Elson

18 Aug	flares	ns
25 Aug	mv **Tannenberg** of Germany	ns

Frederick Edward Crick

30 Oct	trawler, **JAP** (LT245), South Bank	stood by
10 Nov	trawler, **Jamaica** (LT185), South Bank	stood by

1966   *Frederick Edward Crick*

10 Jan	trawler, **Barton Queen** (LT298), Newcome	assisted
11 Jan	trawler, **St. Georges** (LT402), Corton	stood by
25 Jan	trawler, **Warbler** (LT63), Newcome	assisted
27 Mar	drilling rig, **Constellation**, North Sea	ns
10 Apr	trawler, **Ormesby Queen** (LT88), sick man	landed 1
11 May	flares	ns
5 Jun	fb, **Leon Laura** (LT589)	saved 4 + boat
19 Jun	catamaran, **Orlando III**, capsized	towed in
3 Jul	ss **Dunera**, sick passenger	landed 1
24 Jul	yacht, **Chianti**, Newcome	saved 3 + boat
9 Aug	yacht, **Fargo** of Harwich	saved 3 + boat
6 Oct	mv **Spray**, Newcome	stood by
15 Oct	trawler, **Seafarer** (LT448)	saved 5 + boat
20 Oct	trawler, **Seafarer** (LT448)	saved 5 + boat

Elizabeth Elson

1 Nov	*RNLB John F Kennedy*	towed in
10 Nov	mv **Elbeg** of Poland, sick man	landed 1

Frederick Edward Crick

15 Dec	tug, **Kendal** of Rochester, Newcome	landed 8
15 Dec	trawler, **Garibaldi J** (LT358), S roads	towed in

1967   *Elizabeth Elson*

14 Jan   fb, **Mizpah** (FR240), Newcome   saved 2 + boat

                Frederick Edward Crick

10 Mar   mv **Patrick** of Antwerp, Holm   escorted
29 Apr   yacht, **Four Rivers**, Corton   escorted
10 May   fb, **Success** (LT483), Pakefield   saved 2 + boat
13 May   yacht, **Contango**, Newcome   saved 6 + boat
5 Jul   trawler, **Renovate** (BM74)   towed in
13 Jul   swimmer, S beach   ns
19 Aug   yacht, **Sea Mist**, Holm   escorted
7 Oct   fb, **Mary Alice** (LT193), Pakefield   saved 2 + boat
1 Nov   yacht, **Vale**, Newcome   saved 3 + boat
4 Nov   fishing boats in a gale   stood by
30 Nov   parachute down in sea   ns

1968   *Frederick Edward Crick*

20 Feb   flares   ns
29 Feb   flares   ns
17 Mar   m cruiser, **Monaco Philomel**   saved 1 + boat
11 May   dinghy adrift   saved 2 + boat

                Edward & Isabella Irwin

20 May   fb, **Meggies** (LT406), Kessingland   saved 3 + boat
26 Jul   dinghy capsized, Pakefield   ns
9 Aug   dinghy capsized, Kessingland   ns
27 Aug   cries for help, Corton   ns

                Frederick Edward Crick

10 Sep   fb, **Coronation** (LT220), Pakefield   saved 2 + boat
29 Sep   ship on fire   ns
27 Oct   fb, **Happy Sisters** (LT353), Pakefield   ns
15 Dec   man trapped on groyne   ns

Total   176 launches, 145 people and 1 caged bird saved.

Coxswain Harry Burgess, he was in charge of Lowestoft lifeboats from 1947 until 1968. (PLRS collection)

The French trawler **En Avant**. In September 1956 her companion trawler **Les Deux Jeannes** sank on the Newcome Sand. All her crew were rescued by the *Michael Stephens*. The **En Avant** is seen here in Lowestoft harbour. (PLRS collection)

The reserve lifeboat RNLB *Greater London (CS No. 3)* entering Lowestoft harbour having rescued two canoeists and their canoe in August 1957. (PLRS collection)

This small yacht **Knud**, a converted ship's boat, ran ashore on the South beach in August 1958. The *Michael Stephens* stood by but was not needed. (PLRS collection)

The mv **Staniel** of Cardiff whose deck cargo of hay caught fire, was beached at Pakefield in June 1959. The *Michael Stephens* had earlier taken off most of her crew. (PLRS collection

The mv **Ameland** of Gluckstadt grounded on the Holm Sand in May 1961. With a list because of a shifted deck cargo she was escorted into Lowestoft. (PLRS collection)

The French trawler **Pierre Louis** (B.2542) in Lowestoft dry dock after hitting the North extension in August 1962. (PLRS collection)

The Aldeburgh lifeboat RNLB *George and Elizabeth Gow* (ON827) was towed in on two occasions during 1963 following engine trouble. She is seen here in April under tow in the Outer Harbour at Lowestoft. (PLRS collection)

RNLB *Frederick Edward Crick* (ON970), stationed at Lowestoft from 1963 until 1986. She is seen here underway in Lowestoft harbour. (Photo by E Graystone, Lowestoft, from PLRS collection)

The stern trawler **Sailfin** (SN58) ran aground off Lowestoft harbour in February 1965. The *Frederick Edward Crick* stood by for over 12 hours.
The trawler is seen here under tow by the Great Yarmouth tug **Richard Lee Barber**
(PLRS collection)

In July 1966 the *Frederick Edward Crick* went out to the ss **Dunera** to bring ashore a sick schoolgirl. (PLRS collection)

The small trawler **Seafarer** (LT448) was towed to safety twice during October 1966. (PLRS collection)

APPENDIX 1

COXSWAINS

ALBERT SPURGEON 1924 – 1947

He was born in 1880 and joined the lifeboat crew before the turn of the century. In 1924 he was elected to serve as Coxswain in succession to John Swan.

In 1922, as a crew member, he was awarded the RNLI Bronze medal for the service to ss **Hopelyn**. After his appointment as Coxswain he was awarded the RNLI Silver medal in 1927 for the service to the smack **Lily of Devon**. A second Bronze medal followed in 1943 for rescuing the crew of a motor minesweeper ashore on the South beach.

The *Lowestoft Journal* of 25 October 1947 reported that Albert Spurgeon had been involved with the lifeboat for over 45 years and in that time had assisted in rescuing over 600 people. They noted that he had received five RNLI Vellums in addition to his three medals.

He died in 1953 and was buried in Lowestoft cemetery with lifeboatmen acting as bearers.

HARRY THOMAS BURGESS 1947 – 1968

He was born in August 1910 and was, at the time, the youngest east coast Coxswain when he was appointed in 1947, his name having first appeared in the crew lists in about 1928.

All reports about Harry Burgess mark him as a quiet unassuming man reluctant to talk about his work in the lifeboat and from a family with a long association with the sea.

At the time of Harry's death on 3rd May 1979, Jack Mitchley, a longtime friend, wrote, "Harry was a man of quiet courage who was the last of the longshoremen in the lifeboat crew. His death has severed the connections of the Burgess – Rose families with the lifeboat which totalled about 600 years service."

The *Lowestoft Journal* of 4th May 1979 reported that, on being asked his attitude to his work with the lifeboat service, Harry replied, "You go because you think you are helping someone, someone in trouble and somebody has got to do it, have'nt they?"

APPENDIX 2

LIFEBOAT DETAILS

Station boats

Agnes Cross (ON663) 1921 – 1939. 124 launches, 209 saved

Built by SE Saunders, Cowes, Isle of Wight as the *John & Mary Meiklam of Gladswood* at a cost of £8620 for the Gorleston station, being transferred to Lowestoft in late 1921. She was renamed in 1922 and appropriated to the gift of Mrs Agnes F Cross of South Kensington, London.
A wooden motor lifeboat 46ft 9ins x 12ft 10ins, with a 60hp Tyler engine.
She was in the relief fleet from 1939 to 1952 when she was sold out of service. She is last known as the **Wimp** in Aden in 1955.

Michael Stephens (ON838) 1939 – 1963. 142 launches, 110 saved.

Built by JS White, Cowes, Isle of Wight at a cost of £10104. She was the gift of Mrs Louise Stephens of Ewhurst, Surrey.
A wooden motor lifeboat 46ft x 12ft 9ins, with two 40hp Ferry diesel engines.
After service at Lowestoft she served at Exmouth until 1968 and was in the relief fleet until 1976 when she was sold out of service.
She retained her RNLI name and in 1999 took part in celebrations at Poole to mark the 175th Anniversary of the RNLI.
In 2006 she was at Newton Ferrers, Devon.

Frederick Edward Crick (ON970) 1963 – 1987. 253 launches, 106 saved.

Built by JS White, Cowes, Isle of Wight at a cost of £40000 from a legacy from the estate of Mrs FM Crick in memory of her husband.
A wooden motor lifeboat 47ft x 12ft 9ins with two 60hp Gardner diesel engines.
She was replaced at Lowestoft and sold out of service in 1987.
She is noted in 2006 as being restored as a pleasure boat on the river Yonne in France.

Relief boats

During the period covered by this book there were several boats which served at the station to cover absences of the station boat. Details are:

City of Bradford I (ON680)
Built by JS White, Cowes, Isle of Wight at a cost of £12758 in 1923 as the *City of Bradford*. Renamed in 1928. She was funded by the City of Bradford lifeboat fund.
Originally she was at the Humber station and in the relief fleet 1929 to 1952.
A wooden motor lifeboat 45ft x 12ft 6ins with a 80hp Weyburn engine.
She was sold out of service in 1952 and renamed **Hammer**. In 2007 her whereabouts were not known.

JB Proudfoot (ON694)
Built by JS White, Cowes, Isle of Wight at a cost of £7530 in 1924 as the *HF Bailey*. Renamed in 1935. She was funded by a legacy from Mr HF Bailey.
She served at Cromer, Southend-on-Sea and Dover. She was in the relief fleet 1935-41, 1945-47 and 1949-56.
A wooden motor lifeboat 45ft x 12ft 6ins with a 80hp White engine.
She was sold out of service in 1956 and renamed **Anatura** and later **Gramarie**. In 2004 she was reported at Marbella, Spain.

John & Mary Meiklam of Gladswood (ON670)
Built by JS White, Cowes, Isle of Wight at a cost of £10993 in 1923 as the *HF Bailey*. Renamed in 1924. She was funded by a legacy from Mr HF Bailey. She served at Cromer and Gorleston and was in the relief fleet 1939-50.
A wooden motor lifeboat 46ft 6ins x 12ft 9ins with a 80hp Weyburn engine.
She was sold out of service in 1950 and renamed **Pen CW**. In 2006 she was reported on display at the old Gorleston lifeboat house with her name restored to *John & Mary Meiklam of Gladswood*.

Mary Scott (ON691)
Built by JS White, Cowes, Isle of Wight at a cost of £7827 in 1925 from the legacy of Miss M Scott. She served at Southwold 1925 to 1940 and was in the relief fleet until 1953.
A wooden motor lifeboat 46ft 6ins x 12ft 9ins with a 80hp White engine.
She was sold out of service in 1953 and renamed **Atanua**. In 2006 she was reported at Gillingham, Kent.

The Lord Southborough (CS No.1) (ON688)
Built by SE Saunders, Cowes, Isle of Wight at a cost of £8997 in 1924. She was funded by the Civil Service Lifeboat Fund. She served at Margate 1925 to 1951 and was in the relief fleet until 1955.
A wooden motor lifeboat 45ft x 12ft 6ins with a 80hp Weyburn engine.
She was sold out of service in 1955. She was reported in 2006 at Sholing, Southampton with her original name.

EMED (ON705) had one spell of duty but was not launched.
Built by JS White, Cowes, Isle of Wight at a cost of £8700 in 1928. She was funded by the legacies of Mr. Dewhurst, Miss Yates, Mr Barnes and Miss Watkins. She served at Walton-on-the-Naze 1928 to 1953 and was in the relief fleet until 1956.
A wooden motor lifeboat 48ft 6ins x 13ft with two 40hp Weyburn engines.
She was sold out of service in 1955 to Chile and renamed **Capitan Christianson**. She was reported at Valparaiso 1955 to 1998. In 2006 she was on static display in that city.

Greater London (CS No.3)(ON704)
Built by JS White, Cowes, Isle of Wight at a cost of £8668 in 1928. She was funded by the Civil Service Lifeboat Fund. She served at Southend-on-Sea 1928 to 1941 and 1945 to 1955. She was in the relief fleet 1941 to 1945 and 1955 to 1957.
A wooden motor lifeboat 48ft 6ins x 13ft with two 40hp Weyburn engines.
She was sold out of service in 1957 to Chile and renamed **ADES I** and later **Captain Francisco Alvarez**. In 2006 she was reported in a dry dock at Colonia, Montevideo.

Cunard (ON728)
Built by Saunders Roe, Cowes, Isle of Wight at a cost of £8324 in 1930. She was a gift from the Cunard Steamship Co. She served at St. Marys, Isles of Scilly 1930 to 1955 and was in the relief fleet until 1969.
A wooden motor lifeboat 45ft 6ins x 12ft 6ins with two 40hp Weyburn engines.
She was sold out of service in 1969 and renamed **Henry Joy**. In 2006 she was reported at Cornie Island, Killough, N Ireland.

Cecil & Lilian Philpott (ON730)
Built by JS White, Cowes, Isle of Wight at a cost of £7982 in 1930. She was a gift from Mrs L Philpott. She served at Newhaven 1930 to 1959 and was in the relief fleet until 1969.
A wooden motor lifeboat 45ft 6ins x 12ft 6ins with two 40hp Weyburn engines.
She was sold out of service in 1969 and renamed **Stenoa**. In 1999 she was owned by Dr Oliver Dansie and took part in the celebrations at Poole to mark the 175th Anniversary of the RNLI. She was reported in 2005 at Titchmarsh Marina.

Elizabeth Elson (ON713)
Built by JS White, Cowes, Isle of Wight at a cost of £8253 in 1929. She was funded from the legacy of Mr B Elson. She served at Angle, Pembrokeshire 1929 to 1956 and was in the relief fleet until 1968.
A wooden motor lifeboat 45ft 6ins x 12ft 6ins with two 40 hp Weyburn engines.
She was sold out of service in 1968 and renamed **Elizabeth E**. In 2004 she was reported at Sneem, Co. Kerry, Ireland.

Edward & Isabella Irwin (ON778)
Built by Groves & Guttridge, Cowes, Isle of Wight at a cost of £7378 in 1935. She was funded by a legacy from Mrs Irwin. She served at Sunderland 1935 to 1963 and was in the relief fleet until 1969.
A wooden motor lifeboat 46ft x 12ft 9ins with two 40hp Weyburn engines.
She was sold out of service in 1969. In 2006 she was at Vickers Yard, Birkenhead prior to going to the Liverpool Maritime Museum.

INDEX TO LIFEBOATS

Abdy Beauclerk (ON751)	24
AED (ON717)	11
Agnes Cross (ON663)	1-30, 40-44, 83
Baltic (ON665)	22
Cecil & Lilian Philpott (ON730)	61, 77, 86
Charles Burton (ON526)	31
City of Bradford I (ON680)	18, 21, 24, 25, 42, 43, 84
Cunard (ON728)	57, 59, 64, 76, 78, 86
Edward & Isabella Irwin (ON778)	71, 80, 86
Eleanor Brown (ON589)	15
Elizabeth Elson (ON713)	67, 69, 78-80, 86
EMED (ON705)	85
Frederick Edward Crick (ON970)	64-72, 78-80, 83
George & Elizabeth Gow (ON827)	64
Greater London (CS No.3) (ON704)	53-55, 75, 85
J B Proudfoot (ON694)	23, 43, 45, 84
John & Mary Meiklam of Gladswood (ON670)	28, 44, 84
Jorn Hodoshen (Dutch)	36
Jose Neville (ON834)	60
Louisa Heartwell (ON495)	15
Louise Stephens (ON820)	31, 60
Mary Scott (ON691)	16, 32, 37, 46, 48 51, 73, 74, 85
Michael Stephens (ON838)	30-65, 73-78, 83
Neeltje Jacuba (Dutch)	29, 36
Royal Silver Jubilee 1910-1935 (ON780)	22
Ruby & Arthur Reed (ON990)	70
Samarbeta (ON1208)	17
Samuel Plimsoll (ON22)	13
The Lord Southborough (CS No.1) (ON688)	74, 85

INDEX TO VESSELS

trawler	Adrian (LT114)	45
lugger	Adriana I (VL.88)	14, 41
tanker	Adroity (London)	53, 74
ss	Agnes (Haugesund)	40
HMT	Alexandrite	46
tanker	Algol	49
mv	Ameland (Gluckstadt)	61, 77
trawler	Antonia Isabella (O.65)	40
ss	Appledore	31, 44
tug	Armina	52, 74
trawler	Arthur (O.171)	40
ss	Ascania (Wiborg)	43
tug	Barton	23
trawler	Barton Queen (LT298)	79
drifter	Bdellium (FR185)	63
fb	Belle (LT167)	8
fb	Belle (LT320)	53, 74
HMS	Belton	70
barque	Berthon (Norway)	16
drifter	Bezaleel (K27)	17, 42
yacht	Bittern	65
trawler	Blarney Rose (LT1242)	43
drifter	Bob Read (LT2)	41
mv	Borthwick (Glasgow)	76
fb	Boy Colin (LT352)	53
fb	Boy Eric (LT1241)	45
dinghy	Boy Fred	11, 41
fb	Boy George (LT380)	56, 75
m.cruiser	Boy Leslie	66, 78
fb	Boy Reggie (LT104)	15, 41, 42
fb	Boy Roy (LT283)	66, 78
fb	Brenjean (LT86)	77
ss	Brightside (Middlesborough)	37, 46
trawler	Brisbane (GY1281)	38
pleasure bt	Britannia	16
tanker	British Empress	53, 75

barge	Britisher (London)	45
tug	Burton	21
ss	Cambrian Coast (Liverpool)	38, 47
m.cruiser	Careema	77
tug	Cervia	58
barge	Cetus (London)	43
smack	Challenger (LT975)	21, 43
yacht	Chianti	79
yacht	Chums	46
salvage vsl.	Cite de Londres (Hull)	26
lightvessel	Cockle	31
smack	Colinda (LT382)	7, 41
smack	Concord (LT263)	43
rig	Constellation	68, 79
yacht	Contango	69, 80
	Corona	8
fb	Coronation (LT220)	80
lightvessel	Corton	19, 25, 38, 53, 62
fb	Cossack	57
drifter	Covent Garden (LT1258)	46
smack	Crecy (LT226)	46
yacht	Daisy	76
drifter	Dauntless Star (LT371)	73
barge	Davenport (Ipswich)	42
tug	Despatch (Lowestoft)	4, 7, 14, 21, 23
drifter	Diligence (FR917)	2, 8, 40
m.cruiser	Dimcyl	48, 73, 74
trawler	Dinorah (GY1107)	41
fb	Dolly Graham (HL135)	65, 78
fb	Don't Know (YH422)	46
fb	Dot (YH386)	75
fb	Dulcie Doris (LT654)	62, 77
tug	Dundas Cross	54
ss	Dunera	68, 79
ss	Dunleith (Poole)	27, 44
smack	Dusky Queen (LT895)	40

ss	E Rose (Gt. Yarmouth)	9, 41
smack	Ebenezer (LT784)	40
mv	Ecctonia	70
ss	Effra (London)	18, 42
mv	Elbeg (Poland)	79
yacht	Eleanor (Boston)	44
trawler	Elnet (IJM.80)	10, 41
yacht	Elsquir	47
	Emblem	40
ss	Empire Dorritt (Glasgow)	36, 46
trawler	En Avant (Boulogne)	54, 75
ss	Enseign Marie Saint-Germain	34, 45
lugger	Enterprise (LT104)	13, 41
tanker	Esso Ottawa (London)	38, 46
tanker	Esso Parentis (Le Havre)	77
tanker	Esso Stockholm (London)	77
fb	Eva May (LT266)	75
trawler	Excellent (B.915)	27, 44
yacht	Fargo (Harwich)	79
mv	Fiducia II (Holland)	70
mv	Filius (Germany)	78
HMS	Fitzroy	24, 25, 43
trawler	Flag Jack (LT1224)	42
m.cruiser	Fortuna II	78
yacht	Fortyn 2B (Holland)	74
mv	Fountains Abbey	62
yacht	Four Rivers	80
m.cruiser	Francomme	18, 42
HMS	Ganges	39
trawler	Garibaldi J (LT358)	69, 79
yacht	Garlock	75
tug	Gele Zee (Holland)	60
	Gellyburn	41
drifter	Generous (LT1148)	41
tug	George Jewson (Gt. Yarmouth)	1, 21, 25
trawler	Georges Ferges (Boulogne)	60, 77
HMS	Godetia	1, 3, 40
trawler	Godsgenard (O.57)	41

fb	Golden Miller (LT83)	21, 43
HMD	Golden News	45
tug	Gondia	58
drifter	Gowan Bank (BF193)	43
trawler	Grasby (LT267)	75
ss	Greenawn (Goole)	41
mv	Gudveig (Oslo)	60, 77
ss	Gunhild (Copenhagen)	40
fb	Happy Days (LT232)	34, 45
fb	Happy Sisters (LT353)	71, 80
fb	Harnser (YH243)	57, 76
yawl	Helgoland	39, 47
trawler	Helping Hand (LT1239)	38, 46
mv	Henfield	77
ketch	Henrietta (Goole)	40
drifter	Henrietta Spashett (LT82)	77
yacht	Heron (Rochester)	50, 73
m.sweeper	HMMS 106	35, 45
mv	Hoocreek	64, 78
brig	Hope (Hartlepool)	13
ss	Hopelyn (Newcastle)	11, 16, 23, 81
mv	Hudson Firth	52
tug	Imperial (Lowestoft)	23
drifter	Impregnable (LT1118)	14, 41
trawler	Ireland's Eye (GY441)	48, 73
smack	Irene (LT976)	2, 8, 40
trawler	Isle of Wight (H852)	47
smack	Ivan (LT1098)	40
smack	Ivanhoe (LT1271)	42
trawler	Jamaica (LT185)	79
m.boat	James	73
trawler	J A P (LT245)	49, 67, 73, 79
tug	Jean Bart	58
fb	Joan (LT259)	21, 30, 42-44
	Johanna Marie	40
fb	John & George (LT155)	42
trawler	John O'Heugh (HL48)	62, 77

m.launch	Joy	28, 44
fb	Joy (LT337)	45
yacht	Kapella	77
	Kate	41
fb	Kathleen Ann (YH260)	76
tug	Kendal (Rochester)	69, 79
mv	Kentwood	74
trawler	Kestrel (LT1097)	12, 14, 41
fb	Kestrel (LT380)	73
trawler	King Athelstan (LT160)	50, 74
yacht	Knud	57, 76
dumb barge	Leeds Saturn	54, 75
yacht	Leila	78
fb	Leon Laura (LT589)	79
trawler	Les Deux Jeannes (Boulogne)	54, 75
smack	Lily of Devon (LT96)	5, 40, 81
mv	Loa Ronn (Denmark)	46
trawler	Loddon (LT309)	49, 50, 73
drifter	Lord Keith (LT181)	44
tug	Lowestoft (Lowestoft)	1, 8, 26, 27, 38, 48-50
smack	Lucky Hit (LT961)	26, 44
fb	Maggie Jane (LT84)	59, 76
yacht	Magpie (Burnham on Crouch)	43
ss	Mallin (Haugesund)	41
schooner	Marie (Kings Lynn)	3
fb	Marina (LT374)	58, 76
m.cruiser	Marina	75
fb	Marion (LT50)	52, 74
ss	Marion Traber	19, 42
fb	Marjorie (LT379)	19, 42
fb	Mary Alice (LT193)	80
schooner	Mary Ann (Jersey)	41
trawler	Meggies (LT406)	71, 80
yacht	Merry Thought (Newhaven)	44
	MFV 1165	37, 46
fb	Mica (LT683)	9, 41
fb	Mica II (LT330)	71

ss	Mile End (London)	41
fb	Mizpah (FR240)	69, 80
trawler	Mollia (LT123)	52, 74
m.cruiser	Monaco Philomel	71, 80
fb	Monbretia (SH96)	51, 74
trawler	Moreleigh (LT170)	78
mv	Mudo (Holland)	49, 73
dredger	Mudsucker	19, 40
rowing bt.	Myra	40
trawler	Narva (Ostend)	41
mv	National Fighter (Monrovia)	58, 76
tug	Ness Point (Lowestoft)	26, 49, 51
lightvessel	Newarp	31
trawler	Newhaven (LT134)	41
HMT	Niblick	45
trimaran	Nimble Venture (Ipswich)	63, 77
hydroplane	Nine	43
ss	Northam	41
smack	Northern Queen (LT1099)	25, 44, 46
mv	Novian Coast	51
drifter	Ocean Sunbeam (YH344)	49, 73
tanker	Olympic Hill (Monrovia)	78
m.cruiser	Orinthia	43
catamaran	Orlando III	79
trawler	Ormesby Queen (LT88)	68, 79
salvage vsl.	Osprey (Grimsby)	9
m.cruiser	Osprey	78
ss	Pagasitikos (Andros)	21, 42
mv	Patrick (Antwerp)	80
trawler	Pattrick (O.164)	3, 40
fb	Pedro (LT289)	75
lugger	Pet (LT132)	13
HMS	Pickle	39
trawler	Pierre Louis (B.2542)	63, 77
drifter	Pilot Star (KY48)	25, 43
dredger	Pioneer	19
ss	PLM 17 (Rouen)	49, 73

dinghy	Pook	77
trawler	Pradnik (SZN.262)	59, 76
trawler	Purple Heather (LT249)	31, 44
ps	Queen of Kent	23
airship	R33	1, 40
	Ranger	8
fb	Ranger (LT855)	45
trawler	Red Snapper (LT303)	74
trawler	Renovate (BM74)	70, 80
drifter	Reward (LT463)	41
tug	Richard Lee Barber (Yarmouth)	37, 38, 49, 50, 63, 66
canoe	Roaring Boys	55, 75
trawler	Rochester (LT153)	21, 43
ss	Rocquaine (Guernsey)	50, 73
trawler	Roger (N.814)	48, 73
mv	Rose of Lancaster (London)	78
yacht	Roustabout	44
HMD	Rowantree (ex BF199)	34, 45
ss	Royal Crown (Hull)	32, 45
tanker	Rudderman	74
trawler	Sailfin (SN58)	66, 78
smack	San Toy (LT1220)	41
HMS	Saxonia	44
barge	Scone (Rochester)	16, 42
	Scotia	8
yacht	Sea Mist	80
fb	Seafarer (LT448)	68, 79
mv	Seaford (London)	76
yacht	Seagull	44
mv	Seaham (Rotterdam)	51, 74
barge	Servic (London)	16, 42
ss	Sheaf Crown (Newcastle)	47
yacht	Sheenan (Whitby)	74
trawler	Shrublands (LT1223)	12, 41
m.launch	Silver Surf (Southwold)	66, 78
drifter	Silver Wave (BF372)	63, 77
trawler	Silverfish (LT340)	67

tug	Simson	63
mv	Skaansund (Norway)	65
lightvessel	Smith's Knoll	32
dinghy	Snipe	43
mv	Sommen (Sweden)	77
fb	Sonny Boy (LT1214)	17, 30, 42-44
HMS	Spider	4, 7-9, 12, 15
mv	Spray	79
HMS	Squirrel	56
trawler	St. Georges (LT402)	67, 79
ss	St. Helena (Finland)	44
lightvessel	St. Nicholas	25, 31
mv	Staniel (Cardiff)	58, 76
tug	Stentor	63
trawler	Strive (LT766)	41
fb	Success (LT483)	80
yacht	Sujan	61, 77
lightvessel	Sunk	66
HMS	Susette	44
smack	Sweet Home (LT1269)	40
tug	Tactful	21
mv	Tannenberg (Germany)	67, 78
ss	Taunton (Grimsby)	1, 40
fb	Tearaway (LT451)	70
fb	Tempo (YH327)	76
fb	Tennessee (LT146)	41
mv	Tennyson	60
fb	Terry (LT379)	25, 43
yacht	The Fly	65, 78
HMT	Thistle	33, 45
tug	Titan	63
trawler	Tobago (LT182)	78
rowing bt.	Tom Bowling	13, 41
fb	Trade Winds (LT230)	74
smack	Trio (LT1272)	41
smack	Try On (LT176)	41, 43
training vsl.	Turletski (Poland)	56

yacht	Ugly	8, 41
trawler	Unda (LT270)	62
fb	Unity (LT428)	46
fb	Urge (LT282)	26
sloop	Vale	70, 80
tug	Vanquisher	58
ss	Velta (Latvia)	23
trawler	Vestrouwen (Nieuport)	77
fb	Viking (LT11)	46
	Vivo	75
HMS	Wallace	30, 44
trawler	Warbler (LT63)	79
THV	Warden (Harwich)	51
yacht	Water Gipsy (Ipswich)	73
HMS	Wave	56
fb	Wavell (LT381)	50, 74
HMS	Waverley	45
trawler	W E H (LT910)	19, 42
dinghy	Whimbrel	43
fb	White Heather (LT353)	56, 75
yacht	White Lady	53, 75
smack	Wide Awake (LT37)	7, 40
barge	Will Everard (London)	23, 43
trawler	William Leech	27
ss	Willowpool (Hartlepool)	31, 44
mv	Winchester Brook	67
fb	Winston (LT142)	45
trawler	Wisemans (LT382)	67
trawler	Yolande (O.30)	4, 40
fb	Young Paul (LT120)	76
yacht	Ziska (Whitby)	37, 46

INDEX TO PEOPLE

(lb) = connected with the RNLI

Ackworth	Cmdr. H (lb)	55
Aguss	Mr Gary	57
Aitken	Mr Geoffrey	70
Allerton	Mr Walter (lb)	8, 21
Ashdown	Sgt. MHC, RAF	35
Ashton-Stray	Mr C	11, 20
Attridge	Skipper John	50
Ayers	Mr George W (lb)	3
Bacon	Mr CW	50
Bagshaw	Mr	37
Ball	Skipper William	38
Balls	Mr Brian	57, 58
Balls	Mr Edward (lb)	58
Balls	Mr James jun	57
Balls	Mr James sen	57
Balls	Mr T	61
Banks	Captain	9
Batho	Sir Charles	4
Beavers	Mr Albert	56
Beavers	Mr Jack	56
Bishop	Mr RH	60
Blowers	Ldg.AC, RAF	55
Blowers	Mr George	56
Booth	Flt.Lt. RS, RAF	1
Broadley	Sgt. R, RAF	35
Brown	Pilot Offr. VK, RAF	33
Buck	Skipper Alfred	62
Bullen	Mr William	53
Burgess	Cllr. Roy	69
Burgess	Coxswain Harry (lb)	48-72, 82
Burrows	Miss	1, 4
Burwood	Mr John L (lb)	12
Butcher	Mr William C (lb)	23
Button	Mr Charles	50

Capps-Jenner	Mr William (lb)	54, 61, 72
Carr	Mr Frank	22
Case	Mr HG	62
Castella	Mr Frank	62
Catchpole	Mr Peter	65
Chamberlain	Rt. Hon. Sir Austen	10
Chapman	Mr Michael, FRICS	71
Charters	Mr SH	59
Citroën	M.Andre	17
Clarke	Skipper Charles	26
Clarke	Mr Glenn, HMCG	63
Clarke	Sqn.Ldr. N, RAF	63
Clarkson	Mr	59
Clayton	Mr	24
Cleveland	WOII, A, RAF	35
Coble	Skipper S	52
Collins	Sgt. MR, RCAF	33
Collis	Mr Bob	30, 33, 35
Cook	Mr FW (lb)	25
Cooper	Mr William (lb)	14
Creighton	Police Insp. R	57
Cross	Mrs Agnes	1
Cross	Flt.Lt. Arthur, RAF	72
Dawson	Sgt. HC, RCAF	35
Delf	Mr Alan	52
Denyer	Sgt. EC, RAF	35
Dowding	Mr Terry	62
Edmonds	Mr E	26
Edmonds	Mrs P	62
Else	Sgt. JS, RAF	33
Engert	Herr. Edward	18
Evans	Cllr. Arthur	4
Evans	Mr Colin	57
Evans	Mr David	65
Fish	Mr W	26
Fisher	Mr GEC (lb)	2

Gaze	Mr J	52
Gee	Dr. AC	53, 59
Gennery	Mr Jack (lb)	27
Gentry	Mr GAM	7
Gibbons	Mr Peter (lb)	72
Gooch	Mrs A	37
Gooch	Sir Thomas	15
Grantham-Hill	Mr Rodney	51
Green	Mr	15
Griffiths	Mr HF	59
Gurton	Mr DK, HMCG	10
Gwilliam	Captain JG	17
Hailey	Mr	8
Hailey	Mr Richard	65
Hales	Mr Jack	50, 68
Harrison	Mr James	55
Harrols	Sgt. RK, RAF	35
Hermes	Mr CE	24
Hermes	Mr LW	24
Hibbert	Mr Tom	62
Hitter	Mr GW (lb)	25
Hook	Coxswain Robert (lb)	14
Hull	Sgt. RD, RAF	33
Humphrey	Maj. SW	29
Hutchinson	Mr Charles H	28
Hutchinson	Mr Ernest	26
Jensen	Mr F	7
Keith	Mr Billy (lb)	72
Knott	Mr T (lb)	64, 65, 72
Lee	Mr G	26
Lewis	Fl.Sgt. WG, RAF	33
Long	Mr Paul	65
Long	Mr Sam	15
Longmuir	Sgt. Jeff, RAF	72
Lyne	Mr Cyril	62

Lyne	Mr Derek	62
Lyne	Mr Michael	62
Markham	Dr. BH	66, 68, 71
Mayel	Frau. Freda	18
McMeekin	Mr Bruce	62
Mills	Mr H	26
Mobbs	Cllr. W	20
Moore	Mr Tom	17
Morris	Mr Jeff	30
Mottistone	Lord	20
Munday	Mr Graham	26
Munnings	Mr George	18
Neilson	Coxswain (lb)	22
Nobbs	Mr George	17
Parham	Adm. Sir F, RN	56
Parker	Miss Bessie	4
Polkinghorn	Mr J (lb)	61
Poole	Mr Terrence	70
Rentoul	Sir Gervais, MP	11, 12
Rhodes	Captain LA	18
Rickard	Mr B (lb)	65
Robinson	Mr H (lb)	72
Rose	Mr C (lb)	48, 50
Rose	Mr Charles (lb)	22
Rose	Mr George (lb)	2, 31, 48
Rose	Mr Henry G (lb)	3, 25
Rose	Mr Jack sen (lb)	2, 8, 16
Rose	Mr Jack jun (lb)	65, 72
Rose	Mr John (lb)	5
Rose	Mr WN (lb)	13
Runacres	Mr Robert	55
Saunders	Mr J	61
Scott	Mr Cecil, HMCG	57
Scott	Mr Ralph (lb)	5, 11
Seago	Mr Geoffrey	57

Shee	Mr GF (lb)	5
Simmons	Mr Charles	50
Simmons	Police constable	51
Sitco	Mr Joe	65
Slugar	Captain H	58
Soanes	Mr John	7
Somerleyton	Lord	20
Somerleyton	Lady	20, 28, 55
Southby-Tailyour	Maj. Ewan	19
Spashett	Mr F	20
Spinks	Mr W (lb)	14
Spurgeon	Coxswain Albert (lb)	1-39, 81
Spurgeon	Mrs Elizabeth	24
Spurgeon	Mr William sen (lb)	24
Spurgeon	Mr William jun	26
Stanford-Tuck	Sqn.Ldr. RR, RAF	32, 33
Stanner	Mr James	28
Stephenson	Mr Ken	62
Steward	Miss Ethel	28
Stoddard	Mr J (lb)	72
Strathearn	Mr David, HMCG	63
Swan	Mr Arthur JH (lb)	25, 34, 35, 38, 39
Swan	Mr Frank (lb)	65
Swan	Coxswain John (lb)	1, 13, 17, 20
Swan	Mr Thompson	17
Taylor	Mr Sydney (lb)	2, 5, 20, 53, 59
Taylor	Sgt. J, RAF	33
Thorpe	Mr William (lb)	70-72
Threadgold	Sgt. RW, RAF	35
Thurston	Mr Robert	21
Trundle	Gunner II, R, RAF	39
Tuck	Mr V (lb)	61, 72
Tucker	Mr Ernest, HMCG	57
Tucker	Mr EK (lb)	59, 66
Upcraft	Mr	65
Upson	Mr Leslie	62

Vaux	Cmdr (lb)	29, 36
Veness	Mr Arthur	62
Wales	HRH The Prince of	6
Walker	Michael	50
Walton	Mr GR (lb)	65
Ward	Coxswain John (lb)	16
Watson	Dr.	62, 68
Wells	Dr. Anthony	51
West	Skipper Frank	64
West	Skipper George	64
Weston	Mr Arthur, HMCG	61
Wharton	Mr Fred	28
Wollaston-Seago	Mr R (lb)	2
Wooden	Mr Ernest AR	34
Wymer	Mr Henry	58